Financing Corporate Growth in the Renewable Energy Industry

Finanzmärkte und Klimawandel

Herausgegeben von
Dirk Schiereck und Paschen von Flotow

Band 2

Christoph Ettenhuber

Financing Corporate Growth in the Renewable Energy Industry

Bibliographic Information published by the Deutsche Nationalbibliothek
The Deutsche Nationalbibliothek lists this publication in the Deutsche Nationalbibliografie; detailed bibliographic data is available in the internet at http://dnb.d-nb.de.

Zugl.: Darmstadt, Techn. Univ., Diss., 2012

Cover and Photo Design:
© Olaf Gloeckler, Atelier Platen, Friedberg

Library of Congress Cataloging-in-Publication Data

Ettenhuber, Christoph, 1979-
　Financing corporate growth in the renewable energy industry / Christoph Ettenhuber.
　　pages cm. — (Finanzmärkte und Klimawandel, ISSN 2190-3069 ; Band 2)
　ISBN 978-3-631-64420-1
　1. Energy industries—Finance. 2. Renewable energy sources. 3. Energy resources development. I. Title.
HD9502.A2E87 2013
333.79'40681—dc23
　　　　　　　　　　　　　　　　　　　　　　　　2013012138

D 17
ISSN 2190-3069
ISBN 978-3-631-64420-1

© Peter Lang GmbH
Internationaler Verlag der Wissenschaften
Frankfurt am Main 2013
All rights reserved.
PL Academic Research is an Imprint of Peter Lang GmbH.

Peter Lang – Frankfurt am Main · Bern · Bruxelles · New York · Oxford · Warszawa · Wien

All parts of this publication are protected by copyright. Any utilisation outside the strict limits of the copyright law, without the permission of the publisher, is forbidden and liable to prosecution. This applies in particular to reproductions, translations, microfilming, and storage and processing in electronic retrieval systems.

www.peterlang.de

Acknowledgements

I would like to take the opportunity to thank some of the institutions, teachers, colleagues and friends who have guided and supported this project. I am especially grateful to the German Ministry of Education and Research (BMBF) for funding this analysis, which forms part of a wider research project on climate change, financial markets and innovation (CFI).

I would like to thank Dirk Schiereck for his encouragement, constructive feedback and helpful guidance throughout the last three years. I marvel at his ability to multi-task and maintain an open-door policy in the face of a heavy teaching and research schedule. He introduced me to finance and raised my interest in empirical topics during a seminar on Behavioral Finance at the European Business School. I am also indebted to Mark Mietzner, who provided critical orientation in the first months of the project.

I've benefited greatly from discussions with colleagues in the Department of Corporate Finance at Darmstadt. I enjoyed the great atmosphere among postgraduate students to which Anit Deb, Christian Babl, Christian Happ, Daniel Maul, Malte Raudszus, Robert Fraunhoffer and Steffen Meinshausen strongly contributed. Christian Babl and Julian Trillig demonstrated remarkable patience during my early (and sometimes challenging) encounters with econometric software. Both were also kind enough to read and comment on particular chapters in draft. To this end I am also indebted to Jocelyn Evans, who kindly commented on the M&A chapter during the 2011 conference of the Academy of Economics and Finance in Jacksonville (USA), and Eric Duca, who generously provided additional analysis on his related convertible bonds paper. I would also like to thank Adrian Fröhling, Lena, Philip and Carlotta Pecher, Katharina Schultz, Mirko Sedlacek, Tim Zeichhardt, Tobias Schultheiss and Uli Weinspach, who have been great sources of inspiration throughout the process.

Finally, my greatest debt of thanks is to my family: to Janine, for her strength and patience; to Alexander, for his valuable advice; to Katrin and David, for their wisdom and vital encouragement; to my father, Peter, for his unconditional support, and to my mother, Ursula, who would have loved to celebrate the completion of this dissertation.

Preface

Kaum eine Branche erscheint so gut geeignet, die Wertgenerierung durch Unternehmenszusammenschlüsse im internationalen Kontext und unter wechselnden Wettbewerbsbedingungen zu untersuchen, wie der Sektor mit Unternehmen aus dem Bereich der erneuerbaren Energien. Das Wettbewerbsumfeld für Unternehmen ist hier nicht nur spannend, weil der Markt für Technologien in diesem Segment vergleichsweise stark fragmentiert ist. Das Wettbewerbsumfeld war für die Unternehmen durch die Bedienung internationaler Märkte und eines starken Projektcharakters bei der Implementierung immer auch grenzüberschreitend und von großen, kaum diversifizierbaren Einzelrisiken und sich plötzlich ändernden politischen Rahmendaten geprägt. Auch deshalb haben in den letzten Jahren viele Unternehmen im Bereich der erneuerbaren Energien durch Akquisitionen verstärkt versucht, ihr volatiles Geschäft durch vertikale und laterale Erweiterungen zu verstetigen.

Unter solchen Konstellationen erscheinen laterale und vertikale M&A-Transaktionen eigentlich attraktiv und sollten deshalb zu positiven Reaktionen an den Kapitalmärkten führen. Aber passiert das auch? Untersuchungen zum Erfolg internationaler vertikaler M&A-Transaktionen sind bis heute rar, der Kenntnisstand insbesondere zum Erfolg von Akquisitionen im Bereich der erneuerbaren Energien ist noch begrenzter, und die Übertragbarkeit der Evidenz zu den Erfahrungen aus anderen Branchen ist mehr als fraglich. Eine umfassende Analyse der Erfolgsdeterminanten von M&A in dieser Branche ist mir bis dato nicht bekannt.

Die vorliegende Arbeit nimmt sich dieser Forschungslücke mit ebenso viel Liebe zum Detail und höchster Sorgfalt an wie den Fragen der kapitalmarktorientierten Eigenkapitalfinanzierung über Kapitalerhöhungen und Wandelanleihen in diesem Sektor. Ihr primäres Ziel war es, anhand von Marktdaten den Erfolg von internationalen Finanzierungsentscheidungen und M&A-Transaktionen im Bereich der erneuerbaren Energien zu ermitteln und wesentliche Determinanten des Erfolgs zu bestimmen. So wird ein objektiver Kenntnisstand erreicht, auf dessen Basis sich fundierte Handlungsempfehlungen für die Industriepraxis ableiten lassen. Zudem finden sich damit auch für wirtschaftspolitische Entscheidungsträger wichtige Informationen zum Verständnis und zur Bewertung des sich weiter fortsetzenden Konzentrationsprozesses.

Herr Ettenhuber kann die selbst gesetzten Ziele in seiner Dissertationsschrift bestens erfüllen. Die Arbeit enthält viele hoch interessante Resultate und ist so geschrieben, dass es den Lesern große Freude machen wird, sie Seite für Seite bis zum Ende zu studieren. Ich wünsche der Arbeit eine weite Verbreitung.

Professor Dr. Dirk Schiereck

Table of Contents

Acknowledgements V
Preface VII
Table of Contents IX
List of Tables XI
List of Figures XIII
List of Abbreviations XV

1 Introduction 1

2 Financing Constraints in the Cleantech Industry – Theory and Evidence 13
   2.1 Market frictions, market failure and funding gaps 15
   2.2 Firm characteristics and economic development 18
   2.3 Theoretical considerations on capital market frictions 23
   2.4 Empirical evidence on capital market frictions 29
   2.5 Cleantech industry evidence 38
   2.6 Conclusion 42

3 Growth Options, Market Timing and Seasoned Equity Offerings in the Renewable Energy Industry 45
   3.1 Data and methodology 48
   3.2 Logit regression analysis 53
   3.3 Risk dynamics 57
   3.4 Use-of-funds analysis 58
   3.5 Conclusion 64

4 Signaling with Convertible Debt in the Renewable Energy Industry 67
   4.1 Literature review 70
   4.2 Data and methodology 72
   4.3 Industry convertible debt structure 73
   4.4 Short-term announcement effects 75
   4.5 Issuer characteristics and failed signals 77
   4.6 Cross-section analysis 80
   4.7 Risk dynamics and buy-side effects 82
   4.8 Conclusion 85

5	M&A-Success in the Renewable Energy Industry	87
	5.1 Literature review	91
	5.2 Data and methodology	93
	5.3 Short-term announcement effects	97
	5.4 Cross-section analysis	103
	5.5 Conclusion	105
6	Concluding Remarks and Outlook	107
Bibliography		111

List of Tables

Table 1.1:	Capacity development of selected renewable energy carriers 1990–2020e	2
Table 1.2:	Renewable energy-induced direct employment effects	5
Table 1.3:	Major fossil fuel-importing and renewable energy-producing countries	6
Table 2.1:	Key characteristics of the German cleantech industry	22
Table 2.2:	Overview of selected international studies on equity gaps	34
Table 2.3:	Overview of selected German studies on equity gaps	36
Table 2.4:	Equity gap perceptions among venture capital investors	37
Table 3.1:	Descriptive SEO sample characteristics	50
Table 3.2:	Logit regression results	54
Table 3.3:	Estimated SEO probabilities	56
Table 3.4:	Risk changes surrounding SEOs	58
Table 3.5	Impact of SEO proceeds on issuer cash balances	60
Table 3.6:	Impact of SEO proceeds on issuer capital expenditures	62
Table 3.7:	Impact of SEO proceeds on issuer capital structure	63
Table 4.1:	Descriptive convertible debt sample characteristics	73
Table 4.2:	CDO and SEO short-term announcement effects	77
Table 4.3:	CDO and SEO market and firm characteristics at issue	78
Table 4.4:	CDO and SEO financial distress indicators at issue	79
Table 4.5:	Cross-section regression results	81
Table 4.6:	Risk changes surrounding CDOs and SEOs	84
Table 5.1:	Sample M&A transactions by acquirer/target continent of origin	94
Table 5.2:	Sample M&A transaction characteristics	94
Table 5.3:	Short-term announcement effects of M&A transactions	98
Table 5.4:	Short-term announcement effects by renewable energy sub-sector	100
Table 5.5:	Short-term announcement effects by transaction characteristics	102
Table 5.6:	Cross-section regression results	104

List of Figures

Figure 2.1: Cressy's theoretical equity gap 17
Figure 2.2: Survey results of perceived expansion constraints of cleantech companies 39
Figure 3.1: Exemplary SEO timing of Nordex SE 46
Figure 3.2: Logit regression variable illustration 52
Figure 4.1: CDO conversion probability distribution at issue 75
Figure 4.2: Event study estimation period and event window lengths 76
Figure 5.1: Consolidation of renewable energy systems manufacturers by sub-sector 90
Figure 5.2: Short-term announcement effects of M&A transactions 97

List of Abbreviations

Bn	Billion
BNEF	Bloomberg New Energy Finance
BMU	Bundesministerium für Umwelt, Naturschutz und Reaktorsicherheit
BVK	Bundesverband Deutscher Kapitalbeteiligungsgesellschaften
CapEx	Capital expenditures
CAR	Cumulative abnormal return
CAAR	Cumulative average abnormal return
CDO	Convertible debt offering
EEG	Erneuerbare-Energien-Gesetz
EUR	Euro currency
GW	Gigawatt
IPCC	Intergovernmental Panel on Climate Change
KWh	Kilowatt hour
M	Million
MTBV	Market-to-book value
M&A	Mergers and acquisitions
Na	Not available
NEX	WilderHill New Energy Global Innovation Index
OLS	Ordinary least squares
PV	Photovoltaics
R&D	Research and development
REPN	Renewable Energy Network for the 21st Century
RoW	Rest of World
SEO	Seasoned equity offering
SIC	Standard Industrial Classification
SME	Small- and medium-sized enterprise
StEG	Stromeinspeisegesetz
TA	Total assets
TWh	Terawatt hour

U.S.	United States
U.K.	United Kingdom
USD	U.S. Dollar currency
WCX	Wilcoxon test

1 Introduction

The renewable energy industry has seen extensive shifts in industry dynamics in recent years. While the use of wind and hydro energy can be traced back to 200 BC, their role in the 19th and 20th century, when energy consumption dramatically increased, was marginal (Fleming and Probert, 1984). It is only since the beginning of this century that renewable energy carriers offer an increasingly competitive alternative to fossil and nuclear energy sources.

Prior to 1990 renewable energy consisted mainly of hydro power. It accounted for approximately 3% of world energy consumption and displayed little growth potential due to the advanced utilization of favorable geographic locations.[1] Wind energy gathered speed in the United States (U.S.) after the government's introduction of investment incentives (1978) following the oil crisis in 1973.[2] The first large-scale wind energy projects materialized in California, where over 16,000 machines with a total capacity of 1.7 gigawatts (GW) were installed between 1981 and 1990 – a small share of world and even hydro energy production (Righter, 1996). However, in 1990, the U.S. represented almost 80% of the world wind market (Kaldellis and Zafirakis, 2011).

In Europe wind energy grew steadily in the 1980s, albeit from an even lower base. This changed with the disaster at the Ukrainian nuclear plant Chernobyl in 1986, which sparked an energy discussion in Europe and ultimately led to the introduction of a large-scale feed-in tariff system in Germany in 1990. The law (Stromeinspeisegesetz, StEG) required utilities to connect renewable energy producers to the grid and buy any electricity at favorable, but variable and technology-independent rates (65–90% of electricity retail price). This lead to boom in the wind energy sector in Europe and, in particular, Germany, where capacity was added at a factor of 55 in the period 1990-1995 (Kaldellis and

---

[1] Hydro's share of electricity production and capacity is considerably higher. The difference is due to the difficulty of storing electricity and the high production costs of hydro energy. Because ramp-up times are shorter than for fossil or nuclear power plants, hydro is primarily used to meet peak demand.

[2] The Public Utility Regulatory Policies Act, implemented under Jimmy Carter in the U.S. in 1978, is widely regarded as the first form of feed-in tariff. It required utilities to buy energy from qualified producers at "avoided costs", a term which left considerable room for interpretation and eventually led to an asymmetric development of renewable energy across the U.S.

Zafirakis, 2011). Photovoltaic systems were less successful mainly because sys-system costs exceeded tariff income.³

By 2000 wind market shares had changed, with Europe now accounting for 70% of total capacity. This year also marks a pivotal milestone in European renewable energy policy: the replacement of the StEG with a fixed-rate, long-term and technology-dependent feed-in-tariff system in Germany (Erneuerbare-Energien-Gesetz, EEG). Since then, renewable energy capacity has soared (see Table 1.1 for the development in Europe and Germany). In 2009, a directive by the European parliament set binding renewable energy targets for each member country by 2020. These ambitious targets incorporate average annual capacity increases of 25% for biomass, 10% for wind and 17% for photovoltaic systems (Bundesministerium für Umwelt, Naturschutz und Reaktorsicherheit [BMU], 2011).

Table 1.1: Capacity development of selected renewable energy carriers 1990–2020e

This table provides selected statistics on the global development of renewable energy carriers. *CAGR* is the compound annual growth rate over the respective period.

	1990		2000		2004		2009		2020e
	(TWh)	CAGR	(TWh)	CAGR	(TWh)	CAGR	(TWh)	CAGR	(TWh)
Europe									
Biomass	17.3	8.9%	40.5	14.2%	68.9	9.4%	107.9	25.0%	1,258.4
Wind	0.8	39.5%	22.3	27.4%	58.8	17.6%	132.3	9.8%	371.8
PV	0.0	25.9%	0.1	62.7%	0.7	83.8%	14.7	17.4%	85.8
Germany									
Biomass	1.4	12.7%	4.7	20.8%	10.1	24.7%	30.3	4.6%	49.6
Wind	0.1	59.5%	7.6	35.0%	25.1	9.0%	38.6	10.4%	114.8
PV	0.0	51.6%	0.1	71.7%	0.6	63.9%	6.6	19.1%	45.1

Source: Bundesministerium für Umwelt, Naturschutz und Reaktorsicherheit (2011)

---

3 Feed-in tariffs in Europe were subsequently introduced by Switzerland (1991), Italy (1992), Denmark (1993), Luxembourg (1994), Spain (1994), Greece (1994), Sweden (1998), Portugal (1999), Norway (1999), Slovenia (1999), France (2001), Austria (2002), Czech Republic (2002), Finland (2007) and the Netherlands (2011). Countries may have had other renewable energy incentives before or after the introduction of a feed-in tariff. The U.S. and U.K. have so far focused on quota systems, see REPN (2012).

The latest acceleration in renewable energy activity was caused by the nuclear accident at Fukushima, Japan, in 2011. It caused a global review of nuclear energy activity (safety, in particular), slowed down its expansion in many emerging countries and even led to a full exit from it in Germany (by 2022), Belgium (by 2025, conditional) and Switzerland (2034). A significant share of the missing nuclear energy capacity is planned to be replaced by renewable sources.

Despite its rapid development and increasing efficiency, all renewable energy carriers so far remain at a cost disadvantage to conventional fossil or nuclear energy means (Fraunhofer ISE, 2012). Market activity has thus predominantly followed government subsidies and tariffs. The social costs of these market interventions are substantial and raise the question as to why policy-makers have so aggressively supported its development. Most arguments center around one of the following benefits associated with renewable energy development.

First and foremost, renewable energy tends to have a less damaging effect on the environment compared to fossil or nuclear sources of energy. With respect to fossil fuel-based carriers, the majority of scientists now agree that anthropogenic greenhouse gas emissions are very likely to be the main reason for the rise in average temperatures (Intergovernmental Panel on Climate Change [IPCC], 2007). This, in turn, has been linked to an increase in extreme weather events, such as floods and droughts. The social and economic costs to these changes are difficult to estimate, but likely very large.[4] Avoiding such costs by means of reducing greenhouse gas emissions is thus a key goal of renewable energy development.[5] Nuclear energy, on the other hand, incurs almost no greenhouse gas emissions, but the collateral damages caused by accidents may be equally disastrous. Fukushima has provided a vivid example of the environmental risks associated with nuclear energy.[6] Moreover, with nuclear waste representing an unresolved disposal issue, technologies without such technological and environmental challenges offer a distinct advantage.

Second, renewable energy has become an attractive market with significant growth prospects. In fact, many renewable energy policies were designed with a

---

4   A discussion of these costs is beyond the scope of this analysis. For more information see IPCC, 2011.

5   For a contextual discussion of external effects see section 2.3.

6   This represents not only a technology risk, but also a security risk. Due to the magnitude of potential damages, nuclear power plants have been argued to represent targets of terrorist attacks.

view to replicating the high-technology success stories in the U.S., such as in-information technology in "Silicon Valley" or media and biotechnology along Route 128 in Massachusetts (Wüstenhagen and Wübker, 2011). Technology hubs have produced highly successful global companies in key industries, and these contribute considerably to the prosperity of their communities by means of investor returns, taxes and high value-added employment. In the case of Silicon Valley, the emergence and initial success of companies has been linked to a variety of favorable factors: a powerful entrepreneurial culture, close ties between universities and start-up firms, the absence of legal and social barriers, the abundant availability of risk capital, the existence of larger technology corporations from which spin-offs could be made, and the support of the public sector (Lerner, 2009). New technologies typically inspire and enable complementary products and services, which may lead to additional growth and make the region more attractive for highly-skilled employees and entrepreneurs. These so-called knowledge spill-overs (Griliches, 1992) and agglomeration effects (Marshall, 1920) may lead to endogenous growth and create barriers to entry for competing regions and nations. This suggests that there is a first-mover advantage and jump-starting an industry can be beneficial.

With these benefits in mind, policy makers have put in place various support schemes to attract investment and gain "critical mass" in renewable energy business. The appeal of this particular sector is thereby increased by the enormous size of the market it seeks to replace: in 2010, the oil and coal markets were worth USD 2.7 trillion and USD 557 billion, respectively (OPEC, 2011).[7] The largest publicly listed oil and gas companies in the Forbes 2000 ranking commanded an (equity) capitalization of USD 2.5 trillion[8] and generated profits of USD 278 billion in 2011 (Forbes, 2012). The replacement of natural resources with "in-house" developed and manufactured technology offers substantial national and regional wealth potential.

Third, attracting renewable energy companies typically has a positive (at least short-term and direct) impact on employment. Recent estimates indicate that renewable energy currently employs five million people globally (Renewable Energy Network for the 21st Century [REPN], 2012). There are, however, significant differences in job characteristics across technologies. With 2.25m jobs, bioenergy accounted for almost half of all renewable energy employment,

---

7   At an average price of USD 85 per barrel of oil and USD 90 per ton of hard coal.
8   Market prices are as of 28 February 2012.

mainly because growing and harvesting feedstock is very labor intensive. Other technologies have a stronger focus on equipment manufacturing, installation, and project operation.

Table 1.2: Renewable energy-induced direct employment effects
This table provides estimated employment statistics of selected countries and regions.

	Global	China	India	Brazil	U.S.	EU[1]	Germany	Spain	Other
Biomass	750	266	58	na	152	273	51	14	2
Biofuels	1,500	na	na	889	47-160	151	23	2	194
Biogas	230	90	85	na	na	53	51	1.4	na
Geothermal	90	na	na	na	10	53	14	0.6	na
Hydro	40	na	12	na	8	16	7	1.6	1
Solar PV	820	300	112	na	82	268	111	28	60
CSP	40	na	na	na	9	na	2	24	na
Solar Heating	900	800	41	na	9	50	12	10	1
Wind Power	670	150	42	14	75	253	101	55	na
Total	5,000	1,606	350	889	392-505	1,117	372	137	33

Source: Renewable Energy Policy Network for the 21st Century (2012, p. 27), compiled in cooperation with International Renewable Energy Agency and the International Labor Organization's green jobs program.

[1] EU data include Germany and Spain, but are derived from different sources.

The total number of jobs is set to increase, although the recent financial crisis and policy changes have slowed down its development. Overcapacities in the solar and wind industry are likely to dampen growth in the mid-term, and market share changes are likely to cause employment shifts towards Asia (REPN, 2012). However, job creation in renewable energy remains high up on the agenda of many policy makers (International Renewable Energy Agency [IRENA], 2011).

Finally, energy supply risks have always featured heavily in renewable energy policy. Energy is a key production factor of many industries and a key component to consumers' costs. Fluctuations in price and availability of energy may thus have severe adverse effects on investment, mobility and consumer spending. The oil crises in 1973 and 1979 as well as oil price hikes in 2004 and 2008[9] demonstrated the dependency of many countries on moderately priced

---

9 On 11 July 2008 a barrel of Brent oil was traded at an all-time high of USD 147.5.

fossil fuels and the political pressure that can be exerted by restricting others from access to it. It comes as no surprise that the largest importers of fossil fuels (China, United States, Japan, Germany) have sizable renewable energy opera-operations.

Table 1.3: Major fossil fuel-importing and renewable energy-producing countries

This table provides selected statistics on the top ten fossil fuel-importing and renewable energy-producing countries.

Top fossil fuel importers 2009/2010 (million tons)												Top renewable energy producers 2011 (GW)					
Oil[1]		Gas[2,3]		Coal[2]		Wind		PV		Biomass	
USA	510	JPN	99	JPN	187	CHN	62.4	GER	24.8	USA	13.7
CHN	199	GER	83	CHN	157	USA	46.9	ITA	12.8	**GER**	**7.2**
JPN	179	ITA	75	KOR	119	**GER**	**29.1**	JPN	4.9	CHN	4.4
IND	159	USA	74	IND	88	ESP	21.7	ESP	4.5	IND	3.8
KOR	115	FRA	46	TWN	63	IND	16.1	USA	4.0	JPN	3.3
GER	**98**	KOR	43	**GER**	**45**	FRA	6.8	CHN	3.1	ITA	2.1
ITA	80	TUR	37	TUR	27	ITA	6.7	FRA	2.8	ESP	0.8
FRA	72	GBR	37	GBR	26	GBR	6.5	BEL	2.0	na	na
NLD	57	UKR	37	ITA	22	CAN	5.3	CZE	2.0	na	na
ESP	56	ESP	36	MAL	19	POR	4.1	AUS	1.3	na	na
RoW	477	RoW	253	RoW	196	RoW	32.4	RoW	4.9	na	na
Total	2,002	Total	820	Total	949	Total	238	Total	70	Total	72

Source: International Energy Agency (2012), Renewable Energy Policy Network for the 21st Century (2012)

[1] Based on data from 2009.
[2] Based on data from 2010.
[3] In billion cubic meters.

On account of the aforementioned factors, there may be good reason for public investment in renewable energy. However, all measures have been designed with a view to bridge only an initial start-up phase, where research and development costs (R&D) and infrastructure deter private investment. Ideally, public assistance is withdrawn once these companies fledge and are able to compete on their own. For some renewable energy technologies, such as on-shore wind, this pivotal milestone may be close by, while others may still require significant investment and innovation (Popp, 2006).

Governments have recently begun to reduce support for the industry at this critical life-cycle stage. In the European Union, Germany, France and Italy, among others, have reduced or capped their photovoltaic (PV) tariffs, mainly because price declines and capacity expansions have exceeded expectations. In the U.S., support for renewable energy has declined in the presence of abundant shale gas resources. There, concerns have also been voiced regarding the level and effectiveness of government subsidies. The bankruptcies of prominent PV systems manufacturers, such as Solyndra, which received USD 538 million in federal loan guarantees (REPN, 2012), have cast doubt on the long-term competitiveness of the industry and fueled reservations regarding government subsidies. In some cases, rates have been cut for other reasons than competitiveness. The sovereign debt crisis has limited the ability of some governments to sustain funding. Spain, Portugal and Greece cut or suspended photovoltaic tariffs for budget reasons in 2011.

Going forward, it seems that the renewable energy industry is at a pivotal life-cycle stage. The transition from government to private funding may be accompanied by severe challenges. Aggressive growth targets of national action plans are confronted with a decreasing ability of policy-makers to finance such growth. Increasing electricity prices, failing renewable energy companies and infrastructure challenges (electricity grid adjustments, storage technologies) have slowed down the development of the industry. Policy-makers have called on technological innovation to improve the competitiveness, facilitate investment and allow for the retraction of government involvement (REPN, 2012).

One of the key questions associated with this process is the role and capacity of private finance to fund renewable energy projects and companies. The general role of finance and its impact on growth has been subject to considerable academic debate. However, the majority of empirical research, as reviewed by Levine (2005), suggests that there is a strong relationship between financial systems, financial instruments and growth. In particular, he notes that (p.867)

> Theoretical models show that financial instruments, markets and institutions may arise to mitigate the effects of information and transaction costs. In emerging to ameliorate market frictions, financial systems change the incentives and constraints facing economic agents. Thus, financial systems may influence saving rates, investment decisions, technological innovation, and hence long-run growth rates.

In most cases, market frictions have a direct, negative impact on the cost of financing. The classic example of market frictions are adverse selection and moral hazard costs,[10] which require investors to add a risk premium to their opportunity cost of finance. To this end, a better understanding of the specific risks of a project or a group of projects may facilitate the structuring of financing instruments that more effectively mirror their risk-return characteristics and better align incentives among principles and agents. The corresponding decline in financing costs should enable a larger quantity of projects being financed, which in turn may affect production processes and marginal costs (Yelle, 1979).

There are strong indicators that market frictions are not equally distributed across industries: high-tech industries (e.g. biotechnology), for example, have been argued to be particularly affected by information asymmetries (Cressy, 2012). In this field, investors often lack the expertise to evaluate investment projects and, as a result, decide against investing in them. These information asymmetries are often aggravated by a life-cycle component. Young companies in emerging industries have been argued to be stronger exposed to them, because the very nature of novel ideas reduces the number of investors accustomed to it (Carpenter and Petersen, 2002).

As a consequence, Levine (2005) argues that the financing needs of emerging companies change over time and financing systems and instruments need to reflect these changes to sustain growth. This view is also shared by Goldfarb, Kirsch, and Shen (2012), who provide a historical perspective to emerging industry finance. In a case study approach, they show that some emerging industries rely on different capital structures to fund growth. Infrastructure projects in the 19th century in the U.S. were predominantly financed with debt (railroads), while the biotechnology industry had a stronger focus on equity. They also show that the role of intermediaries (e.g. banks, venture capital managers) and institutions (public and private capital markets) change with industry characteristics.

The particularities and characteristics of the renewable energy set it apart from many others. While subsidies have created sizable markets equal to those of mature sectors, renewable energy companies retain some of the risk features and information asymmetry characteristics so common to emerging industries. Moreover, climate externalities, capital intensity and regulatory risks may pose particular financing challenges and impact the industries development.

---

10 See section 2.3 for a discussion on the concepts of adverse selection and moral hazard.

Despite its relevance, corporate finance research on renewable energy companies remains scarce. The majority of existing literature focuses on the indirect impact of policy design (see, for example, Lund, 2009) or the particularities of project finance (Gerhard, Rüschen, and Sandhövel, 2011) in achieving market penetration. Some of the few exceptions include Wüstenhagen and Wübker (2011) and Kenny (2011), who point to financing frictions on the market for renewable energy venture capital, and Chan (2009) who provides initial risk and return characteristics of publicly listed renewable energy companies.

To the best of our knowledge, a detailed analysis of the corporate finance characteristics of renewable energy companies has not been conducted. This may be partly due to data availability, which is a result of the novelty of the industry, the limited number of publicly listed entities and the difficulties in separating it from adjacent disciplines, such as engineering and chemicals. Unfortunately, public market activity of renewable energy companies gathered speed shortly before the financial crisis,[11] which had a strongly negative effect on the public offering activity as a whole (Ritter, 2010).

We seek to fill this gap and divide our analysis in four central research questions. First, we take a look at the fundamental financing constraints of renewable energy companies. We review market failure arguments and discuss recent empirical findings as a foundation to subsequent analyses. As these financing constraints may be shared by other environmentally-focused technologies, we ex-tend our sample to include so-called cleantech companies.[12] To reflect industry structure, we focus on smaller and privately held companies. The analysis also has a tilt towards equity financing, because most authors argue that the risk-return characteristics of equity make it a more efficient instrument to finance emerging industries (Carpenter and Petersen, 2002; Goldfarb, Kirsch, and Shen, 2012).

---

11   The term is frequently used to describe the time period after the bankruptcy of Lehman Brothers Inc. in October 2008.

12   We follow the BMU (2009) in its positive definition of cleantech and categorize any company from the following sub-sectors to belong to this group: renewable energy, re-cycling, sustainable mobility, energy efficiency, water and wastewater management, and resource efficiency. In a similar fashion, we define renewable energy as technologies from the following sub-sectors: solar, wind, ocean, other hydro, biomass, geothermal, biofuels, hydrogen and energy efficiency.

It centers on the question whether there is convincing empirical evidence for the existence of market failure and so-called funding gaps – a necessary, but insufficient condition for direct government intervention in private capital markets (Witt and Hack, 2008). It also seeks to identify which factors and company characteristics increase the likelihood of market failure. We start with a brief discussion of the impact of small- and medium-sized companies (SME) on important socio-economic factors such as growth and employment. It continues to analyze selected international, national (Germany) and industry-specific studies.

The review shows that there is support for the existence of a size/transaction cost-based equity gap for investments ranging from EUR 150,000 to EUR 1,500,000. There are, however, considerable methodical caveats to consider, particularly when estimating capital demand. The cleantech industry's small- and medium-sized structure makes it prone to this transaction cost-based underfunding. The presence of climate externalities, regulatory risks and the industry's capital-intensive nature tend to aggravate the financing problem.

Turning to public capital markets, the second part focuses on the capital raising behavior of listed renewable energy firms. We document the timely clustering of security offerings and ask to what extent asymmetric information, market timing and growth opportunities impact the decision to issue equity. The presence of a widespread market timing motive would seriously question the presence of financing constraints. We use logit regressions and financial statement analysis to investigate 462 seasoned equity offerings in the renewable energy industry over the period 2000 through 2009. Our results are consistent with the notion that market timing plays a less prominent role in this high-growth, emerging industry setting, since company age rather than stock price run-ups and market-to-book ratios explain the issuing decision. However, we do find significant underperformance after the issue, which cannot be explained by risk dynamics. Our use-of-funds analysis shows that the stockpiling of cash is the exception rather than the rule as most additional liquidity is used to finance investment. All these factors point to a capital demand rather than a market timing motive for issuing equity.

Part three advances our analysis to hybrid securities as a response to risk features and capital intensity of the industry. Despite the well-documented problems associated with the use of fixed income securities in risky projects, Chandler (1954) shows that, in the eighteenth century, highly uncertain and capital-intensive infrastructure projects (railroads) were successfully financed with debt. The combination of debt and equity in hybrid securities may thus offer an attractive alternative to overcome the various challenges associated with renewable energy investment. We analyze the structure and use of 44 convertible debt offerings and compare it to 285 seasoned equity securities. Contrary to prior cross-industry research, we show that convertible debt in the renewable energy industry displays a distinct debt-like structure. We show, however, that the preferred security is equity, and that convertible debt is used primarily by companies with high financial distress indicators. We further show that convertible issuers face high adverse selection costs, which they attempt to mitigate by sending signals about firm quality. These signals tend to fail because the market anticipates this behavior, and perceives convertible issuers as being priced out of equity markets.

Finally, we investigate whether business combinations are perceived as a valuable means to renewable energy growth. Mergers and acquisitions (M&A) may provide companies with an improved access to technology and finance, effectively removing previous financial constraints in the common case of larger and more powerful suitors. Share price reactions to such combinations may reveal important information regarding the market's perception of future growth and profitability. Using a data set of 337 M&A transactions conducted between 2000 and 2009, we find positive and significant returns to acquirers and targets. This is remarkable to the extent that most cross-industry studies document zero or negative wealth effects to acquiring shareholders. These have linked the negative returns to the market's anticipation of discretionary spending (Jensen, 1986) and so-called empire building (Roll, 1986) by acquiring firm's management. Cross-sectional regression analysis supports this view by showing that acquirer size and high valuations are negatively correlated with announcement returns. Moreover, we find that acquirers from outside the renewable industry tend to earn positive abnormal returns, which we interpret as a sustainability signal that can be spread over other assets.

In summary, this dissertation provides insights on the financing constraints and financing behavior of the renewable energy industry. Due to its government-induced growth, the size of the addressable market, its industry characteristics and current life-cycle stage, it offers a unique and highly relevant perspective to emerging industry and corporate finance research alike. On an academic level, we contribute to both strands by showing that there is an industry and life-cycle component to security design (and timing). Our results may also provide valuable insights for policy makers, members of the financial services industry and other members of the general public by improving their understanding of the financing dynamics in the renewable energy industry.

2 Financing Constraints in the Cleantech Industry – Theory and Evidence

This chapter seeks to identify and review the financing constraints of cleantech companies. It is motivated by fact that an increasing number of governments directly influence the corporate financing of the industry. A recent example is the public-private partnership fund by the British government, which makes at-arms-length investments in cleantech companies (Department of Business, Innovation and Skills, 2010). In May 2012, the U.S. announced the imposition of antidumping tariffs on Chinese solar panels. China has been alleged to subsidize the industry by means of cheap credit (The New York Times, 2012). The Cleantech Group, a market intelligence provider, notes (2010, p. 5):

> Indeed, it was at the government level the first strong signs of a new 'space race' began emerging, with the prize being dominance of the new cleantech markets. Fueled by unprecedented quantities of "green and clean" stimulus money, cities, states, provinces and countries have aggressively begun competing to grow cleantech businesses, to bring innovation to market, to attract inward investment and to brand themselves as hubs of cleantech growth.

Many governments regard the emerging renewable energy market as an opportunity to grow their economies, relocate value chains and attract high value-added employment. To achieve this, governments' support has grown beyond the provision of infrastructure and fundamental research.

Such interventions are frequently justified by market failure. In some cases and for various reasons, it has been argued, capital markets fail to allocate funds to their most efficient use (European Commission, 2006). Young, small and innovative companies, in particular, would suffer from a lack of financing. Given their above average contribution to innovation, growth and employment, financing constraints would have a substantial impact on the socio-economic development of an economy. To limit welfare losses to society, funding gaps should be attenuated by the government (OECD, 2006).

While many politicians act on the assumption of market failure, academics challenge its very existence. Due to the various potential sources of such financing constraints, and the difficulty in isolating their impact, there have been multiple academic approaches to it (Cressy, 2002). These have frequently produced countervailing views (Levine, 2005; Meza and Webb, 1987; Stiglitz and Weiss, 1981). To assess the current state of research in the field, this analysis reviews theoretical and empirical studies on financing constraints.

Specifically, it asks whether there is convincing theoretical basis and empirical evidence for the existence of a funding gap and which factors and company characteristics stimulate its development. We thereby take an industry approach and focus on cleantech companies. We do so because of the elevated public and political interest in the development of this group of companies, the vast resources that have been used to nurture the industry and the mixed results it has produced so far. The "Boom and Bust" in the (German) solar industry highlights the risks associated with political asset allocation.

Witt and Hack (2008) identify four conditions for the economically justified intervention of governments in private (capital) markets: (i) the positive impact of the market in question on the development of the economy as a whole, (ii) the presence of market failure, (iii) the effective means of governments to mitigate market failure, and (iv) the ability to do so at reasonable costs (efficiency). This paper focuses on the second condition and only briefly touches on the first. There is already a rich body of research evaluating policy responses to market failure, yet very little is known about the fundamental financing constraints. Due to the structure and risk profile of the industry, our analysis has a natural tilt towards small- and medium-sized companies (SME) and equity financing.

Against this background, section 2.1 first establishes important definitions and marks off key concepts. Section 2.2 summarizes the role and impact of SMEs on important macroeconomic indicators such as growth, employment and innovation. It further describes the performance and structural characteristics of the cleantech industry in Germany, which is indeed small- and medium-sized in nature: 74% of all cleantech companies in Germany generated less than EUR 10 million in annual revenues in 2007; 88% generated less than EUR 50 million (BMU, 2010). Our analysis thus applies to a large fraction of cleantech companies in Germany. Section 2.3 addresses the theoretical arguments of market frictions and is divided into demand- and supply-side based factors. Section 2.4 reviews empirical studies in the field, which are identified using a semi-structured query approach in pertinent literature databases.[13] The review outlines popular methods in detecting equity gaps and is divided into international, national and industry-level research.

---

[13] We query the databases Ebscohost, JStor and SSRN with the keywords "funding gap", "equity gap", "financing constraints" and "financial constraints". Searches are also run in Google scholar to identify politics-oriented/initiated studies in the field.

While subject to ample qualifications, the majority of evidence suggests that there are funding constraints for smaller-sized investments between EUR 150,000 and EUR 1,500,000. They are driven primarily by a high fixed component in transaction costs (e.g. auditing, legal and structuring), which render many smaller investments commercially unattractive. Beyond the size criterion, no other factor (e.g. company age or degree of innovation) has convincingly been shown to influence the magnitude or existence of a funding gap. This may be partly due to the methodological difficulties in measuring financial constraints, in particular with respect to estimating capital demand.

The cleantech industry's small- and medium-sized structure makes it prone to size-related underfunding. It is further subject to external climate effects, which reduce the return to investors and, in turn, limit their willingness to invest in the industry. Attempts by policy makers to internalize those effects often come at a cost that many investors are equally unwilling to take: regulatory risk. Changes in feed-in tariffs for renewable energy and uncertainty regarding electricity grid standards, a cornerstone in managing the highly volatile electricity production by renewable energy carriers, are just two primary examples of this risk factor. Finally, the capital intensity and longevity of many cleantech projects do not correspond to the investment criteria of many risk capital pools, which reduces the financing supply to the industry.

2.1 Market frictions, market failure and funding gaps

In a first step we mark off some of the key terms mentioned above. The terms market frictions, market failure and funding gaps are often used rather arbitrarily to describe an elusive financing constraint of single companies or groups of companies. These discussions share the notion that companies are unable to efficiently finance their (otherwise profitable) investment projects. There are, however, differences in the level and severity of the financing constraint.

The term "market failure" goes back to Bator (1958, p. 351) and describes

> [...] the failure of a more or less idealized system of price-market institutions to sustain "desirable" activities or estop [sic] "undesirable" activities. The desirability, in turn, is evaluated relative to the solution values of some explicit or implied maximum-welfare problem.

It is an equilibrium situation where the market fails to allocate resources to their (Pareto) efficient use, i.e. there is another allocation where at least one market participant can improve his situation without impairing the position of others. The source of this failure is often related to non-competitive markets, external effects and public goods. In our context, the financing of renewable energy projects may provide non-priced benefits to society, such as reduced congestion and pollution. In other words, there is a difference between the private and social (net) return of the project, due to which investors may fail to provide funds to the most efficient social use.

One strand of this literature argues in favor of a "gap" (Hubbard, 1998) or a "wedge" (Carpenter and Petersen, 2002) between the internal and external cost of finance. This difference is caused by the effects of informationally opaque markets (non-competitive markets), where outside investors fail to appropriately evaluate project qualities. This leads them to increase interest rates, to compensate for additional risk, or cap funding altogether, which may lead to a situation where there is an excess demand for funding which the market fails to clear (Carpenter and Petersen, 2002).

A key feature of market failure is persistency. Markets regularly adapt to shocks which may cause temporary disturbances in asset valuations and allocations. One example is the frequently reported "over-shooting" of asset prices (Zeira, 1999). Since these are typically not equilibrium situations, but rather short-term market adjustments, they do not represent market failure. However, there is also evidence that these processes can be slow and "sticky". Private equity funds typically require capital commitments over a period of ten years, making them slow to adapt to market shocks (Lerner, 2002a). Such slow market processes may have an effect on the financing of companies by temporarily reducing the supply of finance during times of market shocks or volatile asset prices. We therefore include these temporary effects in our analyses and use either the term market friction or market imperfection to describe them.

The concept of funding gaps is more difficult to isolate. It has been used to describe a special case of market failure which would warrant government intervention (Cressy, 2002). It has also been used to describe credit rationing, i.e. banks' behavior of limiting lending due to an inability to distinguish between project qualities (Evans and Jovanovic, 1989). Cressy (2012, p. 288) defines funding gaps as "[…] as an extreme case of financial market failure where demand for finance is greater than supply and prices and other parameters do not adjust to eliminate the difference." Others have made a similar distinction in that market failure makes financing more expensive, while funding gaps describe situations where there is a complete lack of financing. However, there is no specification as to what constitutes "extreme" (as in Cressy's definition), and any market failure will entail a marginal project for which no funding is available as increased financing costs render it economically unviable.

This figure displays an equity gap model where *SNR* is the social net return, *PNR* the private net return (defined as internal rate of return net of the investor's cost of capital) and C_0 is the initial investment cost (including due diligence). The term *Angel finance* is typically used to describe wealthy individuals who invest their private funds in firms at their very early stage, when the enterprise's capital demand exceeds the resources of the entrepreneur. *Institutional VC* typically describes institutional investor's funds which are managed by (external) fund managers

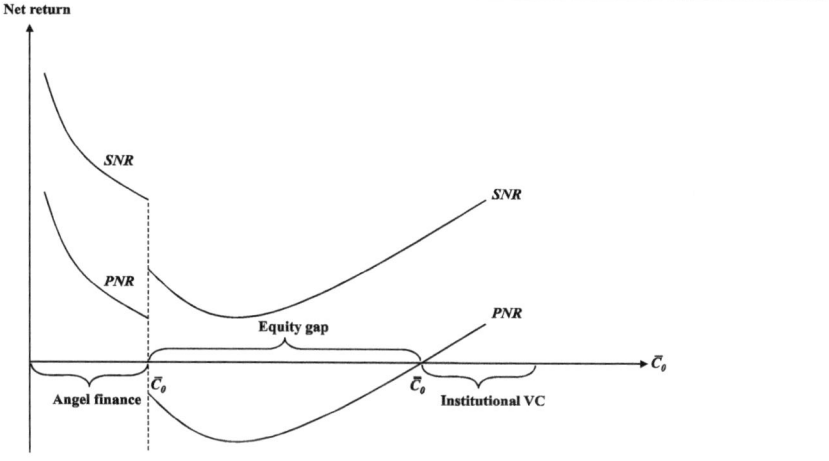

Source: Cressy (2012, p. 283)

Figure 2.1: Cressy's theoretical equity gap

When focusing on equity financing, perceptions as to what constitutes an equity gap remain similarly vague. Harding and Cowling (2006, p. 17) describe the "[…] venture capital market failure that manifests itself in the paradox of an over-supply of capital alongside an inability of small firms to access appropriate amounts and types of growth finance." Lawton (2002, p. 9), following the Bank of England (2000), adds a life-cycle component and defines it as a situation "[…] where the [equity] funding requirements of a company are greater than those that can be met by small-scale providers of finance but not substantially enough to be considered by the large equity providers." In a similar fashion,

Cressy (2012) models an equity gap where market failure and the introduction of transaction costs (at point $\bar{C}_0$ in Figure 2.1) causes the private net return of the investment to be negative, while the net return to society is positive. At investment level $\bar{\bar{C}}_0$ relative transaction costs are small enough to render the private net return positive and facilitate investment. The model contains several simplifications which limit its generality. In particular, projects may exhibit different social and private net return functions (which may never turn positive, for example) and may also not correlate as strongly as graphically suggested.

Based on these considerations, we use the term funding gap to describe a special group of projects which suffer from market failure. These projects exhibit a positive net return to society and negative private net return due to market failure. Some projects may exhibit returns large enough to be financed despite the effects of market failure. From a policy-maker's point of view, the continuum of funding gap projects would start with those that exhibit a high social net return and only a marginally negative private net return and end with projects whose marginal social net return is zero. However, in our analysis we do not restrict the definition to dependent on investment size. It is rather the aim of this chapter to review the dimensions that cause social and private return to deviate.

2.2 Firm characteristics and economic development

An important precondition to our analysis is the beneficial impact of the industry, or group of companies, in question on the socio-economic development of a society. If this is not the case, then any policy responses to mitigate the effect of market failure would be ineffective or even aggravate the problem. Due to our focus on renewable energy, and its predominantly small- and medium-sized structure, we will focus on the impact of these two characteristics on economic growth and employment. However, as these relationships are not at the heart of our analysis, we will restrict it to a brief summary of important results of recent research. For a more detailed discussion see, for example, Praag and Versloot (2008) or Witt and Hack (2008).

2.2.1 Company size, innovation and economic growth

A key driver of economic growth is innovation. Studies by Abramowitz (1956) and Solow (1957) show that economic growth in the United States between 1870 and 1950 is largely due to innovation, since input factor growth accounted for only 15%. Since then, the key role of innovation has been documented by multiple studies with ever increasing data sets (Lerner, 2009). If innovation accounts for the bulk of growth and welfare gains, then the question arises as to where in an economy or by whom those innovations are developed.

The relationship between innovative activity and company size is ambiguous, as studies show both a positive and negative correlation (Lerner, 2009). In most cases, the default hypothesis is that larger companies are less innovative since their structure and culture, set up to maximize output under certain constraints and rules, are inadequate to question and change those rules. It has also been argued that the incentive structures in small and large companies, both intrinsic and extrinsic, are genuinely different. In smaller companies, equity ownership among employees tends to be widespread, and the impact of employees on the performance of the company tends to be high (compared to larger firms). This leads to a higher level of motivation, which fuels the development of new products and services. Larger companies, on the other hand, have more research and development (R&D) resources at their disposal as well as a higher level of process and market expertise (Acs and Gifford, 1996).

Most research approaches resort to R&D expenses, patents and patent citations to measure innovative activity (Acs and Audretsch, 2005). Some also use manager interviews to collect more qualitative information on new products and services. Praag and Versloot (2008) analyze 57 studies published in high-impact journals[14] on the economic contribution of SMEs.[15] They find no difference in R&D spending compared to larger firms. However, they do find differences regarding the quantity and quality of innovation between the two groups. SMEs produce a smaller amount (number of patents) of higher quality (patent citations) innovations, which have been developed more efficiently (patents per

---

14 Articles published in AA- or A-ranked journals according to the Dutch Tinbergen Institute Research School. Additionally, any relevant studies published in the following journals: The Small Business Economics Journal, The Journal of Business Venturing, Strategic Management Journal, Academy of Management Journal and Administrative Science Quarterly.
15 Praag and Versloot (2008) define SMEs as "entrepreneurial firms" with less than 100 employees and a life of less than 7 years.

employee). Regarding the commercialization of innovation, the authors find no significant difference between the groups. Lerner (2009) argues that smaller and younger companies play a key role in the commercialization of innovations, because their structure allows them to exploit market opportunities more quickly. Similarly, smaller companies tend to adopt innovations faster, although to a lesser extent than larger companies. These results suggest that both groups, SMEs and larger entities, make important contributions to innovation, albeit in different segments. Both groups differ with respect to the number of innovations, the efficiency with which they are developed, quality, and the speed with which they adopt innovations. Difficult to measure and widely neglected by the research community is the impact of smaller firms on the innovating activity of larger firms. It seems plausible that the existence of young, innovative companies and the corresponding competitive threat stimulate the innovative activity of larger companies. The previously described segmentation is no contradiction as the studies mainly rely on static analyses that do not follow the evolution of smaller firms.

2.2.2 Company size and employment

The majority of studies document a positive relationship between new ventures and employment. Evidence includes an increasing number of country-level studies such as on the United States (Acs and Mueller, 2008), Germany (Fritsch, 2008), Great Britain (Müller, Stel, and Storey, 2008), Portugal (Baptista, Escária, and Madruga, 2008), and the Netherlands (Stel and Suddle, 2008). The development of employment is typically characterized by a short-term positive effect (approximately one year), a mid-term negative effect (two to five years) and a long-term positive effect. Following Fritsch and Müller (2008), this pattern can be explained by new jobs created by ventures in the short-term, selection and crowding-out effects in the mid-term, and indirect supply-effects, i.e. increased competitiveness, in the long-term. The indirect, long-term competitive effect plays a key role in the analysis as the direct effect of new ventures is relatively small. It remains also the most difficult to estimate.

However, there is reason to believe that the effect is not homogenous. Fritsch (2008), in his literature review, notes that there are significant differences across sectors and regions. Stel, Carree, and Thurik (2005), in their analysis of 36 countries, find a positive impact of new ventures on employment for developed countries, but no – or even negative – effects for developing countries with low start-up rates. Stel and Suddle (2008) find stronger employment effects for the industrial sector compared to the construction, transport, communications and services sectors. Furthermore, stronger effects are shown for regions with high population densities and regions with higher start-up rates (Müller, Stel, and Storey, 2008; Stel and Suddle, 2008). Despite these differences, most researchers agree on the main, positive relationship between new ventures and employment. Praag and Versloot (2008, p. 365) conclude that "The unambiguous results lead to the conclusion that entrepreneurial firms have a disproportionately high contribution to the creation of jobs."

2.2.3 The role and structure of the cleantech industry in Germany

Similar studies on the socio-economic impact of the cleantech industry are available neither on an international nor national level. Most studies focus on the direct growth or employment impact and neglect any potential crowding-out effects. Those studies available, however, point to significant effects on both metrics in Germany. One problem is that many studies define cleantech differently, and mostly opportunistically, depending on the focus of the study or data availability. This problem is aggravated by the fact that cleantech is no separate industry, but rather a cross-section function, which shares many characteristics with engineering and the chemicals industry. Moreover, the reviewed studies typically take a "gross" view, i.e. they do not consider spill-over or competitive effects in other industries, nor do they incorporate the substantial costs to society by means of various support mechanisms.

Many studies on the role of cleantech have been initiated by German government organizations and ministries. They predominantly conclude that increasing environmental challenges will induce a dynamic long-term demand for clean technologies and services. According to Edler et al. (2009), this offers rich growth opportunities for countries with strong industrial base, such as Germany. The most recent, representative cleantech study published by the BMU (2009) divides cleantech into six "lead markets": renewable energy, energy efficiency, resource/material efficiency, recycling, water management and sustainable mobility. German companies in these categories were surveyed using a questionnaire by consultancy firm Roland Berger.[16]

---

16 The sample size amounted to 1,300 companies and 200 research institutions.

The results show that cleantech is already a sizable economic factor today: In 2007, cleantech related economic activity contributed 8% to German GDP, a figure that is estimated to grow to 14% by 2020. The surveyed companies recorded revenue increases in the range of 15% (sustainable mobility, water management) to 29% (renewable energy) over the period 2005 through 2007, well above the German national average. The survey expects that this growth will have accelerated over the period through 2010.

The work forces similarly grew above the national average, at 14%. The variation across sub-sectors, however, was substantial with resource efficiency re-cording only 3% and renewable energy posting a 21% increase in employment. One reason for large direct effects is the high degree of vertical manufacture, which amounted to 65% versus, for example, the 22% of the German automotive industry. German companies are also very innovative: since 2004, the number of cleantech patents grew by 19% annually to reach 1.044 in 2007. Accord-ing to the study, Germany ranks first in the number of global cleantech patents (23%), followed by the U.S. (22%) and Japan (19%).

Table 2.1: Key characteristics of the German cleantech industry

This table provides key results of a survey commissioned by the Bundesministerium für Umwelt, Naturschutz und Reaktorsicherheit and executed by consultant firm Roland Berger. The survey was conducted in October 2008 and includes responses from 1,300 companies and 200 academic institutions.

German cleantech industry by...					
...activity		...Global market share		...annual revenues category	
Planning and consultancy	33%	Renewable energy	~30%	< EUR 10m	74%
Manufacturing	18%	Recycling	~24%	EUR 10–50m	14%
Plant engineering	12%	Sustainable mobility	~18%	> EUR 50m	12%
Trading	10%	Energy efficiency	~12%		
Plant operations (primary)	9%	Water management	~10%		
Contracting	2%	Resource efficiency	~6%		
Plant operations	2%				
Other	14%				

Source: Bundesministerium für Umwelt, Naturschutz und Reaktorsicherheit (2009)

German cleantech companies are mostly small- and medium-sized in nature, fast growing and have a strong base in the services sector (33% of all surveyed companies). In 2007, 88% generated revenues of less than EUR 50 million, 74%

even less than EUR 10 million. The share of very small firms decreased by 6% compared to an earlier version of the survey, which points to a healthy consolidation and increasing development of the industry. Approximately 90% of companies are profitable, although the dispersion is very large, particularly among small companies.

Summarizing this section, we conclude that the precondition to our analysis met. Small- and medium-sized companies have a predominantly positive effect on economic growth and employment. Most (raw) indicators also suggest that the cleantech industry provides a healthy stimulus to the economy and offers significant employment opportunities. We now turn to the sources of financing frictions, which we divide into two groups depending on where they (mainly) originate from.

2.3 Theoretical considerations on capital market frictions

2.3.1 Supply-side based frictions

One of the key arguments for the existence of capital market frictions is the presence of asymmetrically distributed information. Borrowers and investors have different levels of information regarding the risk-return characteristics of the project, which eventually leads to different views regarding the cost of finance.

The borrower is at an information advantage regarding the prospects and risks of the investment. The investor is unable to accurately evaluate the project given the novelty of the idea and the lack of comparable projects. This problem is particularly pronounced in new technologies and industries (Cressy, 2002). The investor's inability to distinguish between profitable and unprofitable projects leads him to require an additional premium for the increased risk in his portfolio. At these increased costs, high quality projects seek funding elsewhere and the investor is left with lower quality projects. This adverse selection of borrowers and deterioration of portfolio quality causes the interest rate to fail as a clearing mechanism on capital markets. Investors have two options to mitigate the effects of asymmetric information: they can either invest in the development of expert knowledge (e.g. industry expertise), which increases transaction costs, or limit funding to cap their risk exposure.

Two things aggravate the asymmetric information problem: first, borrowers have a strong incentive to overemphasize the benefits and play down the risks of the project to minimize financing costs, because these directly impact his residual return. Additionally, many entrepreneurs have been found to be overconfident regarding the prospects of their project, making it even more difficult to evaluate it (Hemer et al., 2006). The presence of asymmetric information has been shown to influence both equity (Greenwald, Stiglitz, and Weiss, 1984; Myers and Majluf, 1984) and debt markets (Stiglitz and Weiss, 1981).

Financial intermediaries such as specialized banks or venture capital investors can help to mitigate the effects of asymmetric information. They regularly evaluate projects and develop industry expertise. They conduct comprehensive pre-investment audits (so-called due diligence), carefully structure transactions to reduce risks (e.g. staggered investments depending on project milestones) and implement extensive control mechanisms to monitor firm performance (Berger and Udell, 1998). These measures reduce risk, but raise transaction costs. As these costs carry a large fix cost component (e.g. a market analysis might be independent of firm size) and tend to increase with the innovative character of the project, transaction costs have a disproportionately effect on small and innovative projects. As a result, small and innovative companies might be particularly affected by the "small-ticket-problem" (Harding, 2000). This has led to many venture capital firms focusing on larger transactions, in particular in countries with advanced financial markets, where increased competition among funds has forced them to decrease costs (H.M. Treasury Small Business Service, 2003).

Another source of capital market frictions are market adjustment processes and behavioral-based effects. The venture capital market is particularly prone to such frictions. These investors often focus on certain industries to develop expert knowledge. Lerner (2002a) argues that venture capital supply is cyclical in nature and asymmetrically distributed among different industries. In 2000, more than 90% of all venture capital investments in the U.S. were made in internet-based or medical technology business models (Lerner, 2002b). A commonly cited example is the simultaneous (venture) financing of 19 hard-disc companies in 1980, of which many failed only shortly thereafter.

Cyclicality and industry concentration can partly be explained by the mid- to long-term fund structures (fund duration, availability of return information), which may lead to an "overshooting" after changes to the capital supply or de-demand of the venture capital industry (Lerner, 2002a). Fund durations are typically not below seven years and it takes another one or two years to "roadshow" potential investors. These long-term capital commitments, together with the illiquid nature of the funds' assets, make adjustments to the capital supply very slow. It may even cause fund volumes to further increase while demand is al-ready in decline, leading to capital supply overshooting demand. This, in turn, may lead to either a capital oversupply with distorted valuations as too many funds are chasing too few deals[17] or to a capital shortage if demand picks up while funds are still in a marketing mode. Fund durations, however, may also prove too short for selective technologies with long development and amortization periods. Energy, in particular, ranks among those industries with the longest lead times, making it a difficult investment area for many venture capitalists.

These market adjustment frictions can be aggravated by herding behavior of financing managers (Devenow and Welch, 1996). Reducing asymmetric information is costly, either by means of increased research spending or the long-term development of industry expertise. One common strategy to reduce these costs is to invest along another fund, which is believed to have more industry expertise. To this end, the well-informed venture fund certifies the company or industry for other investors. Some funds regularly engage in so-called syndicated transactions to fund larger financing rounds and share transaction costs. The uninformed fund thereby typically takes on a minority or junior role. This strategy, however, may increase company-specific risks, because the fund relies on the risk assessment of a third party and the small number of venture investments limits the diversification of such risks. Additionally, if many funds pursue this strategy, the certification function of the better informed fund may lead to a capital oversupply and distorted valuations in some industries and a shortage of capital in others.

A third argument for the existence of capital market frictions is the presence of external effects (externalities). These relate to (Laffont, 2008, p. 16)

> [...] indirect effects of consumption or production activity, that is, effects on agents other than the originator of such activity which do not work through the price system. In a private competitive economy, equilibria [sic] will not be in general

---

17   The term relates to the article "Money Chasing Deals? The Impact of Fund Inflows on Private Equity Valuations" by Gompers and Lerner (2003).

Pareto optimal since they will reflect only private (direct) effects and not social (di-(direct plus indirect) effects of economic activity.

In this case, companies are not (fully) rewarded for socially beneficial action (or fined for socially adverse action), because many aspects of their actions are not captured by contracts or prices. Two types of externalities play a particularly important role for cleantech companies: research and development as well as climate externalities. Market participants typically cannot (fully) capture the effects of their research and development activities. A new, innovative process may be easily copied by competitors; a new product may trigger a range of complementary products, which are not marketed by the originator; and while a higher employment level certainly benefits a society as a whole, it may not benefit the innovator himself. Anticipating this inability to capture the (full) returns of his work, he may choose not to invest and undertake the project at all. For the society as a whole, this leads to a suboptimal level of investment (Griliches, 1992).

Obviously, this problem is attenuated by the use of patents. However, not all innovations are patentable, nor is it always economically efficient to do so. Patents may be in conflict across different countries and regions – a fact that runs counter to the increasingly global nature of markets. Even if the innovator holds a title to a certain product, service or technology, he may not be able to enforce those rights. In fact, some companies choose not to patent major technologies because of their inability and costs associated to enforce them (Levin et al., 1987).

In addition to research and development externalities, cleantech companies are particularly affected by climate externalities. These relate to the consumption of public goods, for example air or water, for which the public is not compensated, and which provide a cost advantage to the originator. Cleantech companies often reduce or substitute these costs but, given that they are not priced, are not (fully) rewarded for this more efficient use of public goods. One of the classic examples is industrial pollution, whose costs to society have long not been priced (Knight, 2010). The European Union tries to internalize some of these external costs with the introduction of tradable emission permits.[18]

---

[18] The U.S. and the EU have established a common framework to assess externalities in energy production. A series of research projects have been launched under the project name "Extern E". For further information see Extern E (2004).

Finally, extensive subsidization of cleantech companies around the world suggests that many governments embrace the presence of agglomeration effects. These are efficiency gains caused by the spatial proximity and the sharing of resources by market participants. First described by Marshall (1920), agglomeration effects originate from knowledge transfer as well as employment and resource synergies and can help to explain the geographic distribution of industries. It may be beneficial to locate a company close by its competitors to share a large pool of specialized workforce and easily exchange industry information. At the same time, it may be advantageous for employees to relocate to a place where employer options are vast. Agglomeration effects can lead to endogenic growth and create barriers to entry, which may help to explain the concept's popularity with policy makers. It suggests that, in the case of industry dynamics, there is a first-mover advantage and jump-starting an industry can be beneficial. Classic examples include the U.S. furniture industry, which is heavily dependent on its access to timber and geographically concentrated in Western North Carolina, despite more advantageous locations elsewhere in the U.S.; or the information technology industry which, despite its independence of tangible resources, is predominantly located in Silicon Valley in California (Lerner, 2009; Strange, 2009).

2.3.2 Demand side-based frictions

Financing problems of SMEs may also be caused by demand-side based frictions. These primarily represent violations of the assumption of rational and utility-maximizing behavior of its economic agents.

Many actions of entrepreneurs seem to be in conflict with the utility-maximizing assumption of economic agents. For example, one of the main reasons to start a business is the desire to be independent (Oakey, 2003). External financing, from a bank overdraft to straight equity, may be regarded as a first step to losing control of this independence. Many entrepreneurs thus intentionally decide not to undertake a project if external financing is required, even if the project has a positive net present value and would increase his or her net wealth (Fraser, 2004). This not only applies to the desire of control, but also to leisure time and living at a suboptimal production location. The motivation to maximize value decreases as soon as the entrepreneur has reached a satisfactory level of income (Oakey, 2007).[19]

In addition to these intentional decisions not to rely on external financing, entrepreneurs may also simply lack the knowledge of where to find financing and how to apply for it. Financiers regularly report financing requests that lack critical information or do not appropriately address key aspects of the business idea (so-called "investment un-readiness"). Financing shortages may also materialize if entrepreneurs do not exploit the financing options available (Achleitner and Poech, 2004; Paul et al., 2005). This may arise if borrowers have reservations against selected types of financing, perceive (market standard) financing costs as prohibitively expensive or simply have too little knowledge about the financing options available. Some authors argue that many entrepreneurs have too little experience in dealing with financiers, which leads to communication problems and a subsequent shortage of financing (Wolf, 2006).

Fraser (2004) conducts a comprehensive survey among British entrepreneurs and finds that 8% of financing requests were never made because the applicants expected to be declined. According to Oakey (2007), this latent capital demand is never made public due to the applicants' fear of a loss of face (and the effort associated with the application). It is of course difficult to evaluate how many of these financing requests would have warranted funding. It seems plausible that the average quality of these projects is below those that were actually made. However, this difficult to quantify part of requests leads to a systematic underestimation of capital demand (Oakey, 2007).

Finally, Pleschak and Werner (1998) and Wolf and Ossenkopf (2005) argue that financial constraints on entrepreneurs may arise from the inefficient use of existing funds. Revenues from non-core activities and cash-saving measures would often alleviate liquidity pressures. Financing problems are thus partially caused by insufficient management skills. This notion is also supported by a growing body of literature on "bootstrapping", i.e. "[…] finding creative ways to avoid the need for external financing through reducing overall cost of operation, improving cash flow, or using financial sources internal to the company" (Ebben and Johnson, 2006, p. 851), which argues that financial constraints cause small firms to operate more efficiently.

---

19   Please note that this behavior does not necessarily constitute a market friction, because the entrepreneur is maximizing his utility.

2.4 Empirical evidence on capital market frictions
2.4.1 Overview

Evidence on the link between one of the aforementioned arguments and financing constraints is scarce to non-existent as researchers struggle to find support for market failure itself. Previous research provides ambiguous results. Stiglitz and Weiss (1981) argue theoretically that the presence of asymmetric information causes firms to be debt constrained. Meza and Webb (1987) show that a slight variation of the Stiglitz and Weiss (1981) model leads to overinvestment, and suggest increased taxing to arrive at a socially optimum level of investment. Cressy (2002), summarizing the results of a symposium on funding gaps, concludes that there is no consensus regarding the existence, size and location of a funding gap. Witt and Hack (2008) draw a similar conclusion after an extensive literature review on venture financing. They qualify their conclusion by noting that there may be subgroups, such as growth or high-tech ventures, which may indeed suffer from underinvestment. Cressy (2002, p. F9) makes a similar exception to his analysis in observing that "One area where many writers feel asymmetric information is at its most damaging to fund-raising is therefore that of young, small high-tech firms, commonly known as Technology-based Small Firms."

Many fundamental studies target the debt capital market to detect SMEs' limited access to financing (Hubbard, 1998).[20] Carpenter and Petersen (2002), for a large sample of U.S. firms, show that small high-tech firms have very limited access to debt financing. They argue that debt is an ill-suited financing instrument for small high-tech firms due to the following reasons: first, these firms have fundamentally different return distributions compared to fixed-income instruments. They are typically skewed,[21] with a high probability of failure, but also high returns in case of success. Debt providers typically do not participate in the large up-side potential of these companies and are left with the high probability of failure, which often leads them to step back from the financing. Second, information asymmetries in high-tech industries tend to be

---

20 There is an extensive body of literature on the merits of the cash flow-sensitivity of investment. An in-depth discussion of these arguments is beyond the scope of this analysis. For further information see Fazzari, Hubbard, and Petersen (1988), Hubbard (1998), Kaplan and Zingales (1997), Almeida, Campello, and Weisbach (2004).
21 For limited liability companies, losses are capped at zero while potential gains are unlim-ited, leading to an asymmetric return distribution.

particularly pronounced and can lead to adverse selection and credit rationing. Third, lenders anticipate an adverse change in behavior after committing debt funds to the borrower (moral hazard). In the presence of debt (and rising with the level of it) equity holders have an incentive to substitute high-risk for low-risk projects (Green, 1984). Since equity can be viewed as a call option with a strike price equal to the face value of debt, and option values increasing with volatility, stock holders have an incentive to invest in particularly risky, even negative net present value projects at the expense of bondholders (so-called risk shifting).[22] Lenders typically respond to this with credit rationing, the use of covenants and contractually limiting certain borrower behavior. Fourth, assets of high-tech companies tend to be highly company-specific and offer little collateral value to lenders in the case of default. Finally, Cornell and Shapiro (1988) argue that financial distress costs are particularly pronounced at small high-tech firms. Future growth options represent a large part of overall firm value and these rapidly depreciate in the presence of financial distress. In this situation, lenders are reluctant to provide extensive debt financing.

Some of these barriers to lending are overcome by entrepreneurs that pledge collateral from their private wealth. To this end, they signal quality and overcome some of the information asymmetry, but also take on substantial personal risks.[23] However, it is the equity market that most authors regard as vital for SME development, and frictions on this market are most likely to impact the development of key technologies (Carpenter and Petersen, 2002; Lerner, 2002a). The easy access to external equity financing in the U.S., both via highly developed public capital markets and the venture capital industry, has been linked to the success of U.S. companies in various high-technology industries (Trester, 1998).

Venture capital in particular has been the subject of various studies. At first sight, this may seem surprising given the scarcity of information about venture capital funds as well as the small fraction of overall SME financing they provide (1.8% of U.S. SME financing according to Berger and Udell (1998), 5% of U.K. SME financing according to Mason and Harrison (1999)). The main reason for this is that venture capital-backed companies are over-represented among companies conducting initial public offerings (IPO) and are generally

---

22 The idea to view equity as an option was first developed by Merton (1974).
23 Relaxing the assumption of the high-technology focus of SMEs shows that debt capital plays an important role in SME financing, as do trade credit and informal investors (Berger and Udell, 1998).

considered highly successful (Lerner, 2009). Some authors argue that this small number of successful companies accounts for the bulk of growth and employment effects to society (Mason and Harrison, 1999). Frictions on the venture capital market would thus have a large impact on society and negatively affect some of the most promising companies in an economy. Venture capital typically provides more than capital to a particularly risky group of companies: fund managers tend to advise on the strategy of the company, help setting up critical processes as the company grows, arrange financing and offer a wide network of business contacts (Gompers and Lerner, 2001). Venture capital can certify a company for other investors, both for debt and equity financing, and can have a multiplier effect with respect to additional financing for the company (more equity capital typically increases its capacity).

2.4.2 International equity gap studies

Focusing on venture capital as a proxy for external risk capital, studies can be classified in three groups (Venturelli and Gualandri, 2009): the first seeks to identify regional disparities in the supply of venture capital (Achleitner et al., 2009; Martin et al., 2005). These studies compare venture capital investments with regional or national economic indicators such as the number of companies or gross domestic product. They are based on the assumption that venture capital demand is driven by these (crude) proxies and that the relationship is linear in nature. If the ratio for a particular region falls below an (arbitrarily) defined threshold, the authors conclude that the region is short of venture capital funding, which in turn impedes its economic development. The basic problem with this approach is that few venture capital investments can be the result of a limited number of investment opportunities. In this case it is not supply, but rather demand that holds back investment. The fact that only a fraction of SMEs is financed by venture capital makes the approach prone to misinterpretation. The regional clustering of industries and the industry focus of many venture capital funds complicate the analysis and raises the question whether gross domestic product or number of corporations are appropriate indicators for venture capital demand.

A second group of studies uses qualitative analysis in form of interviews and surveys (Blanchflower and Oswald, 1998; Harding and Cowling, 2006; North, Smallbone, and Vickers, 2001; Wolf, 2006). These include surveys of both capital supply (venture capital managers, banks, public financing institutions) and demand (entrepreneurs of various industries, the general public).

Unfortunately, the results tend to vary greatly with respect to panel composition (Harding and Cowling, 2006). Moreover, behavioral biases may render the results unreliable. Additionally, some investors, such as informal investors and so-called Business Angels, are difficult to identify because they are not organized. These studies also suffer from the general problem that investment projects are heterogeneous in nature. The rejection of a financing request may be due to market failure, but may also stem from the financier questioning the prospects of the project. Reliable evidence with this approach would require two identical investment projects, evaluated by unbiased investors, and resulting in different outcomes.

A third group estimates funding gaps with quantitative demand analysis (Harding and Cowling, 2006; Venturelli and Gualandri, 2009). The latter estimate the capital demand of SMEs in the Italian Emilia Romagna region. Based on financial statement data and with an ample set of assumptions regarding operating and financing decisions,[24] the authors estimate the financing demand in relation to the growth of these companies. Their result is an estimate of the equity capital demand, which allows no direct conclusion regarding an equity gap, but offers valuable insights into the capital demand distribution of SMEs. Due to the amount and quality of information required, quantitative demand analysis is the least frequently used method.

Mindful of these methodological caveats, most studies of the U.K., the most frequently researched market, estimate an equity gap for investments between GBP 250,000 and 1,500,000. The most recent study by Harding and Cowling (2006) also points to a gap of GBP 10,000 to 30,000 for companies between 18 and 24 months of age to meet regulatory and administrative cost. Bannock Consulting (2001), in a study for the European Commission, and based on surveys and expert interviews, quantifies a funding gap of USD 250,000 to 5,000,000 for the U.S. The authors argue that funding gaps in Europe should be more pronounced given the smaller pool of risk capital in Europe.

Table 2.2 provides an overview and key results of studies that quantify the equity demand or an equity gap. It is almost unanimously explained by transaction costs. The fixed cost component of identifying, analyzing, structuring and monitoring a transaction renders smaller transactions unattractive for many investors. Evidence on other influence factors, such as the

---

24 For example, the model requires constant leverage and capital intensity. These assumptions weigh heavily on the explanatory power of the model.

level of innovative activity, age, or growth, remains ambiguous at best and is often based on single case studies (Wolf, 2006).

Despite the difficulties in identifying and quantifying a gap, many governments and politicians act on the assumption of its existence. The European Commission published an indirect measure of an equity gap by setting limits (up to EUR 1,000,000) to government funding of private companies (European Commission, 2001). In 2006, the Commission lifted this number to EUR 1,500,000 per year explaining (European Commission, 2006, p. 4):

> While it is the primary role of the market to provide sufficient risk capital in the Community, there is an 'equity gap' in the risk capital market, a persistent capital market imperfection preventing supply from meeting demand at a price acceptable to both sides, which negatively affects European SMEs. The gap concerns mainly high-tech innovative and mostly young firms with high growth potential. However, a wider range of firms of different ages and in different sectors with smaller growth potential that cannot find financing for their expansion projects without external risk capital may also be affected. The existence of the equity gap may justify the granting of State aid in certain limited circumstances. If properly targeted, State aid in support of risk capital provision can be an effective means to alleviate the identified market failures in this field and to leverage private capital.

This view is also shared outside the European Union. In a survey by the Organization for Economic Co-operation and Development (OECD), 80% of OECD countries and 90% of non-OECD countries assumed the existence of a funding gap in their respective countries. Non-OECD countries thereby believed that gaps primarily existed on the debt market, while OECD countries expected frictions on the equity market.

Table 2.2: Overview of selected international studies on equity gaps

The table provides details on selected international studies on equity gaps. Studies are identified using a semi-structured query approach in pertinent databases. Only studies that quantify an equity gap are included. Qualitative analysis denotes survey- or interview-based evidence. Quantitative studies include macro (fund flows, investments) and micro (firm-level financial statement) analyses.

Study	Published in	Type of study	Sample	Key results
Venturelli and Gualandri (2009)	Journal of Small Business and Enterprise Development	Quantitative demand analysis	1,458 industrial and service companies (less than EUR 50m in revenues, growing) of the Emilia Romagna region in Italy.	Short-term liabilities are main financing source of SMEs; average annual equity demand of EUR 147,300; equity demand decreases with age and size; no correlation of equity demand with innovative activity.
SQW Consulting (2009)	Contract research for the British Government (Department of Business, Innovation and Skills)	Qualitative supply analysis	VC and Business Angel investments in the U.K. ('98–'07).	Equity gap for investments between GBP 250,000 and 2,000,000; for R&D and capital-intensive companies up to GBP 15,000,000.
Harding and Cowling (2006)	Journal of Small Business and Enterprise Development	Qualitative and quantitative demand analysis	60 semi-structured expert interviews (U.K.), Global Entrepreneurship Monitor (U.K., '01–'03), Competitiveness in Banking Review (U.K., '99), The Work Foundation's Enterprise Panel of Inquiry (U.K., '03).	Equity gaps for investments between GBP 150,000 and 1,500,000 as well as GBP 10,000 and 30,000; upper bound rises as financial market develops (competition forces funds to focus on larger deals).
Mason and Harrison (2003)	Regional studies	Quantitative supply analysis; regional venture capital ratios	VC investments of members of the British Venture Capital Association ('95–'01).	Equity gap for investments between GBP 250,000 and 1,000,000; regional differences in VC investments.
Bannock Consulting (2001)	Contract research for the European Commission (DG Enterprise)	Qualitative supply analysis	Expert interviews with public and hybrid financing institutions in the European Union ('99–'00).	Equity gap for 5,000 innovative companies and hybrid (mezzanine) financing gap for approx. 200,000 companies for investments between EUR 100,000 and 1,000,000.

2.4.3 Germany-focused equity gap studies

Similar quantifications of an equity gap for Germany do not exist. Studies on the German market tend to be more qualitative in nature and focus on the regional availability of venture capital (see Table 2.3 for an overview). They center on the question whether venture capital fund managers have a regional preference for investments, i.e. tend to invest close to the location of the fund's management, either to reduce transaction and monitoring costs or out of a behavioral bias (Cumming and Johan, 2006; Powell et al., 2002). If there was such a preference, regions outside the sphere of major cities, where offices are typically located, would economically suffer from a lack of risk capital. Fritsch and Schilder (2007) survey 75 early-stage investors in Germany and conclude that spatial proximity has no effect on fund investments and the regional availability of risk capital. The design of the study, however, may include an incentive bias. Even if managers were aware of a behavioral bias, they would most likely not agree to it given that investors in the fund are unlikely to reward such a behavior.

Martin et al. (2005) conduct a comparative study on the regional availability of risk capital in U.K. and Germany. They add an analysis of the regional dispersion of venture capital investments to a survey of both public and private venture capital investors. They show that funds in Germany are concentrated in six major German cities (Munich, Frankfurt, Düsseldorf, Hannover, Hamburg, and Berlin) and that they predominantly invest in their respective federal state: over the period 1998 through 2000, Munich funds made 68.3% of their investments in Bavaria, Hannover funds 53.3% in Lower Saxony and 45.7% of Berlin-based funds remained in the German capital. Martin et al. (2005) also calculate venture capital ratios for each federal state (state share of venture capital investments divided by state share of number of firms, relative to total number of venture capital and number of firms in Germany) and define a ratio below one as an indication for a venture capital shortage and above one as relative over-supply. With this metric, Berlin, Bavaria, Hamburg and Baden-Württemberg were host to many venture capital-receiving companies. There are many difficulties in interpreting this simple indicator, in particular the structural differences of city states (Berlin and Hamburg) and territorial states (Bavaria and Baden-Württemberg) as well as temporary fluctuations in this metric. However, the authors conclude that there is a regional, local preference which is (partly) caused by lower transaction and monitoring costs.

Table 2.3: *Overview of selected German studies on equity gaps*

The table provides details on selected German studies on equity gaps. Studies are identified using a semi-structured query approach in pertinent databases. Qualitative analysis denotes survey- or interview-based evidence. Quantitative studies include macro (fund flows, investments) and micro (firm-level financial statement) analyses.

Study	Published in	Type of study	Sample	Key results
Achleitner et al. (2009)	International Journal of Entrepreneurship & Innovation Management	Quantitative supply analysis; regional venture capital ratios	Data from the German venture capital association (BVK) 2004–2006, Zukunftsatlas 2007 data.	Potential VC gap dependent on indicator (volume vs. number) and investment phase (early-stage vs. growth); Brandenburg with the lowest VC level; public VC provide more advice than private VC
Fritsch and Schilder (2007)	Empirical Entrepreneurship in Europe: New Perspectives	Qualitative supply analysis	Semi-structured interviews of 75 investors and managers (22 VC funds, 11 business angels, 23 banks, 17 bank-owned and 12 public funds).	No regional preference regarding investments; syndication reduces monitoring costs.
Wolf (2006)	Working paper series Unternehmen und Region – Fraunhofer ISI	Qualitative demand analysis	Surveys of 554 companies that received (start-up) state aid ('01); 687 service sector start-ups des ('99).	No correlation between perceived funding gap and company age, size or level of innovative activity. Management quality and financing environment critical for perceived shortage of funds over time.
Schilder (2006)	Freiberger working papers - Technische Universität Bergakademie Freiberg	Qualitative supply analysis	23 public VC funds (KFW, MBG, funds of savings banks (Sparkassen)), 28 private VC funds.	Public and private venture capital funds address different segments; public venture capitalists provide less advice than private venture capitalists.
Martin et al. (2005)	Environment and Planning A	Quantitative and qualitative supply analysis; regional venture capital ratios	BVK- and KFW-data ('99–'01); interviews with 107 German and 60 British VC funds.	Regional investment bias of VC managers; availability of VC can stimulate economic development; strongest perception of an equity gap (among VC managers) of up to EUR 750,000.

Surprisingly, in this survey, it is not the high-tech segment that is most affected by a lack of equity capital, but the low tech manufacturing segment, which encompasses large parts of the German "Mittelstand". Oakey (2007) in his analysis of "non-stellar" SMEs argues that these companies do not exhibit the challenging growth and profitability characteristics that many venture capital investors require.

Table 2.4: *Equity gap perceptions among venture capital investors*

This table provides selected results of a survey (2005) among German (107) and U.K. (60) venture capital investors.

(in %, multiple responses)	**Germany**	**U.K.**
Regional gaps		
Most frequently mentioned as affected by equity gaps	13.2 Brandenburg	8.1 North West
	12.0 Sachsen	8.0 South West
Sector gaps		
Information technology and media	18.8	31.6
Other services	23.5	10.5
Manufacturing high-tech	17.6	15.8
Manufacturing low-tech	40.0	13.2
Life sciences	12.9	26.3
Stage gaps		
Seed stage	52.6	54.5
Start-up	38.1	56.4
Expansion	14.4	7.3
MBO/MBI[1]	4.1	1.8
Replacement	3.1	3.6
Bridge	3.1	5.5
Deal-size gaps		
< € 150,000	42.9	42.9
€ 150,000 – € 375,000	48.4	51.8
€ 375,000 – € 750,000	26.4	48.2
€ 750,000 – € 1.5m	13.2	35.7
€ 1.5m – € 5.0m	12.1	21.4
€ 5.0m – € 50.0m	4.4	1.8

Source: Martin et al. (2005, p. 1222)

[1] MBO/MBI: Management buy-out/ management buy-in.

Without access to this pool of risk capital and having grown out of government support schemes, these companies' (slower) growth would be seriously impeded by the lack of capital. Largely consistent with previously discussed size characteristics, investors estimate funding gaps to be severe in the category up to EUR 750,000.

Achleitner et al. (2009) also calculate venture capital ratios for each German federal state. Their data are derived from the German venture capital association (Bundesverband Deutscher Kapitalbeteiligungsgesellschaften, BVK) and allow for a differentiation of start-up and growth investments (for both volume and number of transactions). Similarly to Martin et al. (2005), they identify Brandenburg as receiving particularly little venture and growth investment. The states of Hessen, Bremen, and North Rhine-Westphalia also record a relatively low level of investment (three out of four ratios below their arbitrarily defined threshold of 0.67). They could not find support for their main hypothesis that the economic development in East and West Germany is related to the availability of venture capital.

Wolf (2006) focuses on the capital demand and analyzes surveys among entrepreneurs. His first data set comprises completed questionnaires of 554 companies located in East Germany, which, at the time of the survey, were between one and five years of age and had previously received public support for their venture (or were located in one of the publicly supported start-up centers). His second survey comprises 687 questionnaires of service sector ventures located across Germany.[25] Wolf (2006) finds no support for a relationship between financing constraints and company characteristics (size, age or innovative activity). In fact, his analysis suggests that management quality, financing preparation and the financing environment have a stronger impact on the availability of funds. Moreover, he finds that the higher the variety of financing instruments used by the company, the lower are the perceived financing constraints. This implies that management and financing know-how rather than structural funding gaps drives the (perceived) financing constraints.

2.5 Cleantech industry evidence

Similar analyses on an industry level are scarce to non-existent. Industry effects only play a subordinate role in the previously reviewed studies. This is surprising given that the sources of financing constraints should feature differently across industries (e.g. asymmetric information, climate externalities).

---

25   The survey was conducted in 1999 among 4.500 service sector ventures.

Some authors pool industries in an often elusive technology-based or high-technology cluster and argue in favor of differences regarding the level of asymmetric in-formation (Carpenter and Petersen, 2002; Oakey, 2003). In one of the few single industry studies, Cressy (2012) surveys 41 executives of U.K. biotechnology companies. He concludes that (p. 288) "[...] the results of this study suggest strongly that difficulties in raising finance have a strong sectoral component and are greater the earlier the stage of the business." Based on the emerging nature of the cleantech industry and its high score with other potential sources of capital market imperfections, we expect cleantech companies to be subject to severe financial constraints.

There is a myriad of research on policy responses to funding gaps in the cleantech sector, yet very little is available on the actual financing constraints of these firms. Most of the research in the field has been initiated by policy-makers and advisory firms. The financial services firm Grant Thornton conducts an annual survey among a large number of privately held companies worldwide. Figure 2.2 shows these firms' view on business expansion constraints in 2010.

This figure displays survey results of the Grant Thornton International Business Monitor 2010. According to the authors, the survey was conducted among 7,400 privately held companies from 36 economies. The cleantech sub-sample includes responses from 160 companies defined as businesses with more than 40% of activity related to cleantech. The figures represent the percentage of businesses rating the respective constraint 4 or 5 on a scale of 1 to 5 where 1 is not a constraint and 5 is a major constraint.

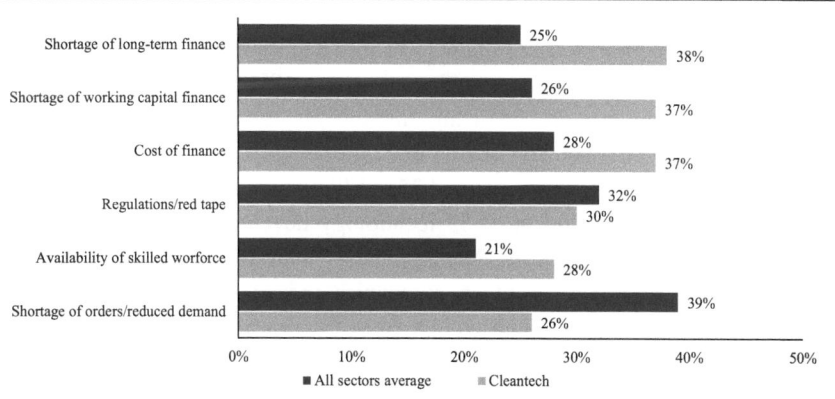

Source: Grant Thornton International Ltd. (2010)

Figure 2.2: Survey results of perceived expansion constraints of cleantech companies

Compared to other industries, cleantech firms perceive financing as a stronger impediment to growth. Almost 40% of cleantech companies see the availability and cost of financing as a key constraint. Demand for cleantech products and services is high, while there are smaller differences regarding regulation and workforce availability. In a similar fashion, Bloomberg New Energy Finance and The Clean Energy Group, the latter a non-profit U.S. organization, conduct 60 open-ended interviews with clean energy entrepreneurs, project developers, investors and policymakers (Bloomberg New Energy Finance, 2010). Interviewees identify two "valleys of death", i.e. two development stages, where the lack of financing impedes the development of cleantech innovation. The so-called technology gap materializes after the new technology has been developed and the project has reached a capital demand level beyond the resources of the initiator. At this stage, projects are often too small or immature for institutional risk capital. The second gap is perceived at the commercialization stage, where risk capital is replaced by larger volume asset development capital, such as project finance. However, due to the capital intensity of the cleantech sector, venture capital investors often need to exit early, when technology risk remains too prominent for other forms of finance. While the participants regard the technology gap as a critical, but manageable problem, the commercialization gap causes more severe financing frictions, which the private finance sector is unable to attenuate. The authors even suggest that (Bloomberg New Energy Finance, 2010, p. 8) "No existing class of financing institutions is effectively positioned to address this particular risk-return profile."

One of the most comprehensive studies was initiated by the European Union and executed by an international research consortium (FUNDETEC, 2007). They analyze public and private financing instruments targeted at environmental technologies and conduct semi-structured interviews. In total, their data set contains interviews with 40 private and 35 public financing experts as well as 33 technology developers. 20 out of 33 technology developers call financing the most important constraint in the development of their company. This leads the authors to conclude that cleantech companies indeed suffer from a lack of funding. They support this with results from interviews with financiers, who perceive cleantech companies as more risky, both in terms of regulatory as well as technology risk. They also argue that many financiers lack the required knowledge to value cleantech projects and that the small- and mid-sized structure of the cleantech industry makes it prone to the small ticket problem.

Cleantech companies would also be at a disadvantage to conventional fossil and nuclear energy sources resulting from external effects.

On the demand side, their results support the view that entrepreneurs have incomplete knowledge about the financing options available and perceive them as being excessively complex to understand. The loss of control among entrepreneurs also featured heavily in the survey results. The entrepreneurs perceived funding gaps to be most severe in the pre-commercialization phase, when the company has grown beyond basic government support and private funds, but remains too small for other forms of commercial funding.

Research on the supply of risk capital in the cleantech industry focuses on interviews with specialized venture capital funds. Wüstenhagen and Teppo (2006) conduct 23 semi-structured interviews with cleantech venture capital investors (11 independent VCs, 11 corporate VCs, one government VC). Their analysis centers on the question why so little venture capital is invested in the renewable energy sector. Interviewees identify several industry-specific risk-factors, which the authors conclude remain yet insufficient to explain the vast differences between renewable energy market size and venture capital investment. Among other factors, cleantech VCs mentioned the increased market adoption risk due to the gatekeeper function of utilities. Investors also bewail the presence of regulatory risks and the lack of liquid exit markets.

Ghosh and Nanda (2010, p. 2) conduct "[…] discussions with several leading venture capital investors" and use case studies to show that cleantech is a challenging field for venture capital investors. They argue that the capital intensity of many cleantech businesses render them unattractive for venture funds. Their return distributions are typically skewed, with few very successful investments compensating for a large fraction of failed projects. Benefits from diversification are large, as additional investments increase the chance of investing in a successful portfolio company. As such, less capital-intensive projects are often favored by venture capital funds, putting cleantech at a disadvantage. They further argue that cleantech investments suffer from a lack of skilled, entrepreneurial managers, illiquid exit markets and elevated regulatory risk, all making cleantech less attractive to venture capital investors. Interestingly, they also show that cleantech venture investment is highly concentrated among few venture capital firms, indicating a high level of asymmetric information and the require-ment of expert knowledge.

Knight (2010) surveys 34 specialized venture capital managers and cleantech companies in the U.S. and the U.K. He argues that renewable energy carriers require local customization, both environmentally and regulatory, which puts them at a scale disadvantage to conventional energy sources. He adds that a higher level of asymmetric information, climate externalities and regulatory risks negatively affects the financing of cleantech companies. His interviews document that the capital intensity renders many cleantech projects unattractive for venture capitalists. He concludes that in order to alleviate the financing problems, regulators must mitigate asymmetric information and internalize external effects.

2.6 Conclusion

This section reviewed the theoretical arguments and empirical evidence on the financing constraints of cleantech companies. Financing is regularly perceived as a key business development constraint. Small- and medium-sized companies, the dominant size category in the cleantech sector, are often particularly affected. Theoretical reasons can be classified in supply-side and demand side-based frictions. These are commonly related to the presence of asymmetric information, transaction costs and climate as well as research externalities. Some borrowers also lack knowledge about financing options or decide intentionally against external financing.

Without pledging personal collateral by the entrepreneur, many SMEs have limited access to debt capital. Their risk-return profile often forces them to rely on external equity capital, which is provided by either institutional (VC, business angels, support program) or informal sources (friends, family). VC-funded companies have been argued to be particularly successful, because VCs typically initiate comprehensive selection processes and provide additional services (e.g. networking, company processes, strategy reviews) to monitor and develop portfolio companies. Many studies thus focus on venture capital markets to proxy frictions in the wider equity capital market.

Measuring such frictions, however, remains methodologically difficult. A reliable test would include estimates of both capital supply and demand. Since investment projects are heterogeneous and risky in nature, it is difficult to differentiate projects that are deemed commercially unsound and those that are subject to market frictions. Researchers either resort to surveys, the relative availability of risk capital (per country, region, GDP) or quantitative demand analyses. All three methodologies have major drawbacks: survey results depend heavily on sample composition and participants' incentives; the regional or sectoral availability of venture capital may be due to a lack of investment opportunities; quantitative demand analyses' ample assumptions and qualifications raise questions about their results' viability. It is therefore the combination of all three methods that should provide the basis for conclusions regarding the existence and size of a gap.

The preponderance of studies consistently documents a funding gap for smaller investments in the range of EUR 150,000 to EUR 1,500,000, in some cases up to five million. It originates from the large fixed costs component of transaction costs in analyzing and structuring a transaction and typically occurs after the business has grown beyond the financing ability of the entrepreneur, but remains too small for institutional venture capital. There are further strong indicators for equity financing constraints of SMEs that do not meet the aggressive growth and profitability requirements of venture capital funds. This is a significant number given that more than 95% of financing requests are declined. Other factors, such as the level of innovation or company age have not convincingly been shown to impact the existence or size of the equity gap. There is further evidence that the size of the equity gap is not stable over time. It seems to be dependent on the maturity of the VC industry and the economic cycle.

While there is no research quantifying a funding gap, there is good reason to believe that financial constraints feature prominently in the cleantech industry. Companies perceive financing as difficult to come by and prohibitively expensive. Due to its small- and medium-sized structure, the cleantech industry is particularly affected by the transaction cost-based funding gap. There is also convincing survey-based evidence that it is not a particularly attractive target industry for venture capitalists. The capital intensity and longevity of many cleantech projects, together with the high level of regulatory risk, render it unattractive for many venture capital funds. The lack of liquid exit markets and the presence of climate externalities aggravate the problem. Potential mitigations include syndication of venture capital investments as well as stable, long-term regulation to minimize the regulatory risk to investors.

There are ample opportunities for future research. There is a lack of studies that quantify equity and debt gaps both on a regional and industry level. The majority of studies target the U.K. or the Italian market. Specifically, the increasing electronic availability of private company accounts in Germany should provide a rich data basis for SME financing insights. There is also a need for more quantitative demand research to avoid the biases and challenges of survey-based evidence.

Parts of this chapter have been published as follows: Ettenuber, C., Schiereck D., & von Flotow, P. (2011). Finanzierungsrestriktionen bei kleinen und mittleren Unternehmen der Umwelttechnologiebranche–Stand der Forschung und offene Fragen. *Zeitschrift für Umweltpolitik und Umweltrecht*, 34(1), 43-72.

3 Growth Options, Market Timing and Seasoned Equity Offerings in the Renewable Energy Industry

Seasoned equity offerings (SEOs) are typically accompanied by previous stock price run-ups, high share valuations and poor stock price performance following the offering (Ritter, 2003). These patterns are inconsistent with two key financial policies: the pecking-order theory (Myers and Majluf, 1984) and various tax and leverage cost trade-off models. According to the pecking-order theory, external equity financing ranks last among the financing choices when other forms are available. Stock price run-ups, however, indicate additional borrowing capacity and thus suggest additional debt rather than equity financing. Trade-off models similarly predict additional leverage (or increased dividends) to return to the firm's target capital structure after the stock price run-up has tilted the balance towards equity (DeAngelo, DeAngelo, and Stulz, 2010).

Given the failure of these theoretical models to explain issuing patterns, Loughran and Ritter (1995) suggest that managers, led by the presence of asymmetric information, time the market and sell equity when valuations are high, providing incumbent shareholders with an advantage at the expense of outside investors.

Many studies have documented market timing for both primary and follow-on equity offerings.[26] Recent research by Howe and Zhang (2010) on SEO cycles over a 33-year period tests prominent drivers of SEO activity and finds support only for the market timing and capital demand hypotheses. While research has produced rich evidence of the very existence of market timing, its economic impact has received less attention. DeAngelo, DeAngelo, and Stulz (2010), in their study of 4,291 industrial SEOs in the U.S. (1973-2001), find support for market timing, but conclude that its impact on the SEO decision is limited: "The problem for the market-timing explanation is that, paraphrasing Sherlock

---

26 Restricting the review to SEOs, issues have been documented to coincide with high valuations (Asquith and Mullins, 1986; Kim and Weisbach, 2008; Korajczyk, Lucas, and McDonald, 1991), negative short-term announcement effects (Masulis and Korwar, 1986; Mikkelson and Partch, 1986), and long-run negative performance (Eckbo, Masulis, and Norli, 2000; Jegadeesh, 2000). There is also a growing body of research using duration analysis to analyze the decision to conduct an SEO (Harjoto and Garen, 2003).

Holmes, many "dogs don't bark" at times when, according to the theory, they should be barking" (p. 276). Specifically, their model suggests that the probability to conduct an SEO increases by only 5.1% (to a total of 8%) when poor market timing indicators are replaced by very favorable indicators. They find support for their life-cycle hypothesis in which young companies with an existing need for capital issue equity to fund growth.

This figure illustrates Nordex SE's share price development and timing of seasoned equity offerings. We consider only transactions that provide liquidity to the company (two purely secondary transactions took place in 2010). Share prices are rebased to 100 at 1 January 2004.

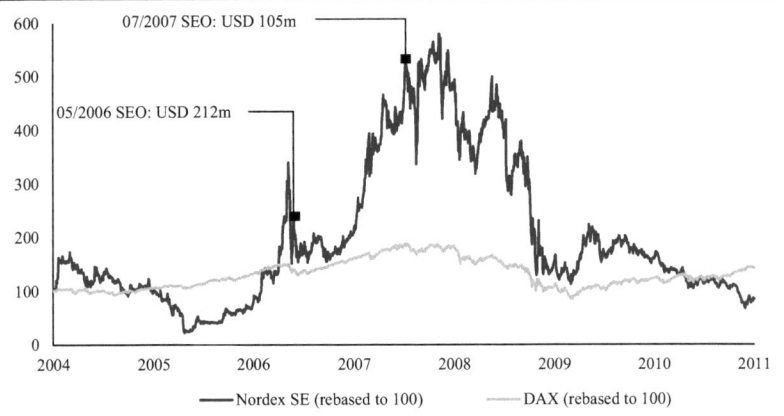

Source: Datastream, Thomson Reuters

Figure 3.1: Exemplary SEO timing of Nordex SE

We investigate the impact of market timing in the renewable energy industry, and our analysis is motivated by two factors that are widely neglected in previous studies: first, broad samples implicitly assume that there are no industry or life-cycle effects. They assume, for example, that market timing played a similar role in information technology and manufacturing at the beginning of the century. Second, in terms of industry life-cycle, they assume that the market timing motive in information technology did not change between the beginning of the century and today.

There are strong indicators that suggest that this might not be the case. Variations in SEO activity have been linked to the notion of "windows of opportunity" (Bayless and Chaplinsky, 1996) and "hot" markets (Ibbotson and Jaffe, 1975). In line with Myers and Majluf's (1984) pecking-order theory, these periods of favorable issuing environment have typically been explained by a de-decrease in asymmetric information, which reduces the adverse pricing effect and thus reduces the overall cost of the equity issue. Firms in an emerging, growing industry may find it easier to signal the presence of strong investment opportunities (e.g. through high capital spending) than firms from mature industries. Bay-less and Chaplinsky (1996) note that "the influence of firm characteristics on announcement date prediction errors suggests that hot and cold markets could result from the clustering of certain types of firms in high and low volume periods" (p. 254). Based on these considerations, we believe that industry and life-cycle effects play a key role in the SEO decisions and valuation patterns should mirror these differences.

We extend the evidence on the effect of market timing in equity issues by concentrating on an emerging industry where the SEO probability is perceived to be much higher than in cross-industry means. The higher this probability the more expected an SEO should be, and expected SEOs may have severe problems to time the market. We expect capital demand and growth opportunities rather than market timing to be key drivers of the SEO decision in this setting. In line with this notion, we expect issuers to utilize their increased equity levels to step up debt financing and spend heavily on capital expenditures (CapEx) to exploit growth opportunities.

The renewable energy industry offers a unique environment to test this hypothesis. The energy industry has been argued to be at a structural change, with alter-native/renewable energy being a high-growth, capital-intensive industry. Both IPO and SEO activity has picked up considerably over the last decade (see Table 3.1 on p. 48). We use a set of 462 international renewable energy SEOs over the period 2000 through 2009. We closely follow DeAngelo, DeAngelo, and Stulz (2010) in their approach and use logit regressions and a use-of-funds-analysis to estimate the impact of market timing on the decision to conduct an SEO.

We show that the market-to-book value (MTBV) and the stock's prior net-of-market returns play a less prominent role in explaining SEO behavior than in most cross-industry studies. Both their statistical and economic significance are weak to non-existent, in line with the notion that market timing is less pronounced in a growth setting. On the other hand, we also find significantly negative future returns with high marginal probabilities, which are not driven by risk changes as suggested by Carlson, Fisher, and Giammarino (2006). All indicators together point to a different issuing behavior of renewable energy issuers. Our company age indicator provides evidence that there is a life-cycle component, because young firms are more likely to issue equity than mature ones.

Our results from the use-of-funds analysis show that the stockpiling of cash is the exception rather than the rule among renewable energy issuers. Most firms would have run out of cash in the year following the SEO, and more than two thirds of our sample would maintain significantly reduced cash balances. Renewable energy issuers spend heavily on CapEx, with 40% spending twice or more of the issue proceeds in the three years surrounding the SEO. Most of them also considerably step up debt financing after the issue. All these indicators suggest that pure discretionary spending of timed proceeds is unlikely.

The contributions of this section are the following: to the best of the authors' knowledge, there is no study directly incorporating a life-cycle and industry impact on the decision to issue seasoned equity. This is surprising to the extent that these effects may have a substantial impact on SEO pricing and valuation. It further offers insights into factors that influence the timely clustering of SEOs and adds to a better understanding of the emergence of "hot" and "cold" markets. Moreover, our analysis represents an indirect test of financing constraints often associated with the renewable energy industry. The presence of a widespread market timing motive would seriously question the existence of such constraints, at least for publicly listed companies. Finally, our focus on an international sample extends the predominantly U.S.-centered evidence on SEO market timing.[27]

The analysis proceeds as follows: subsection 3.1 describes our data selection and methodology. Subsections 3.2 and 3.3 present the results of the logit regression and an important robustness test, respectively. We include the latter because any significant risk changes surrounding the issue would render our results unreliable. Subsection 3.4 outlines the results of our complimentary use-of-funds-analysis. Subsection 3.5 concludes.

3.1 Data and methodology

In the absence of a standard industrial code (SIC) for renewable energy firms, we collect issuer names from two alternative sources: we include all firms in the

---

[27] Outside the U.S., there have been few international (Kim and Weisbach, 2008) and national studies (Bo, Huang, and Wang, 2011; Cohen, Papadaki, and Siougle, 2007).

Bloomberg New Energy Finance (BNEF) database as of March 2010. BNEF is one of the most comprehensive databases in the field of renewable energy and carbon markets and includes many small and young issuers. We further add any of the 100 companies included in the WilderHill New Energy Global Innovation Index (NEX, also administered by Bloomberg) as of the same date. NEX companies tend to be larger in size and more mature than BNEF companies. We combine the sub-sector classifications of both sources. We run these issuer names in the Thomson Reuters database to identify SEO transactions and rely on Datastream for share price and financial statement data.

We include only issues that match the following sampling conditions: we require issues to have taken place between 2000 and 2009, to be made in ordinary shares only (we exclude any restricted, special rights or derivative securities such as convertibles or Global/American Depository Receipts, unit offerings or closed-end funds), to have proceeds information available (primary and secondary shares), and to be larger than USD 1 million in size. We choose this rather low threshold because of the critical (binary) condition of event versus non-event for our logit regression. Throughout our analysis, "proceeds" refers to cash received by the firm and we exclude any purely secondary issues, where only existing stockholders sell shares in the company. Table 3.1 provides descriptive statistics on our sample of 462 SEOs.

The sample exhibits considerable heterogeneity in issue size and geographic as well as sub-sector dispersion. While the number of SEOs has almost consistently increased over the ten-year period, IPO activity offers a differentiated view on open financing windows and is clustered around the 2004-2007 period. The distribution of proceeds is skewed, with a few large issues accounting for the bulk of the overall cash raised: the largest ten percent of SEOs provide almost 65% of total proceeds. The large number of small issues primarily stems from companies domiciled in Australia, the United Kingdom and, to a lesser extent, the United States, Canada and New Zealand, where stock exchanges offer entry level segments with low firm size requirements. Many Australian, U.S., and Canadian firms issue small amounts more than once a year, often to finance short-term working capital needs.

Table 3.1: Descriptive SEO sample characteristics

Descriptive statistics on 462 seasoned equity offerings by Bloomberg New Energy Finance (BNEF)/ WilderHill New Energy Global Innovation Index (NEX) renewable energy firms over the period 2000–2009 and data available on Thomson Reuters. Issuer nationality follows Thomson Reuters, industry classification follows BNEF/NEX. *Proceeds* relate to cash received by the firm, net of any purely secondary proceeds. *Average primary* share is the ratio of cash received by the firm to total issue proceeds.

	Number of SEOs	Proceeds (USDm)	Mean proceeds (USDm)	Median proceeds (USDm)	Average primary share	Number of IPOs
2000	13	2,532	194.8	69.9	97.2%	9
2001	9	453	50.3	38.6	99.9%	4
2002	12	318	26.5	4.6	100.0%	8
2003	20	617	30.9	4.6	94.3%	5
2004	20	566	28.3	5.4	100.0%	14
2005	21	782	37.2	25.2	100.0%	29
2006	50	2,600	52.0	21.6	97.1%	29
2007	92	8,903	96.8	21.3	94.9%	23
2008	93	4,939	53.1	15.0	96.5%	4
2009	132	9,000	68.2	12.2	97.8%	2
Total/All SEOs	**462**	**30,711**	**66**	**15**	**97.0%**	**127**
United States	146	10,527	72	28	93.8%	23
Australia	86	826	10	5	100.0%	14
United Kingdom	60	935	16	4	99.1%	22
Canada	39	962	25	19	96.0%	7
Germany	28	1,763	63	33	95.1%	16
Taiwan	17	637	37	14	100.0%	7
Japan	9	1,175	131	48	94.0%	2
China	9	539	60	16	88.9%	3
New Zealand	7	83	12	6	100.0%	2
Hong Kong	7	1,341	192	30	96.4%	3
Rest of World	54	11,923	220.8	41	98.8%	28
Total/All SEOs	**462**	**30,711**	**66**	**15**	**97.0%**	**127**
Efficiency	90	4,286	48	15	95.4%	16
Solar	86	7,238	84	30	95.5%	31
Wind	57	11,905	209	48	97.8%	15
Biofuels	53	2,443	46	12	99.7%	18
Geothermal	37	825	22	5	96.8%	9
Power Storage	36	1,043	29	12	95.5%	8
Biomass Waste	34	904	27	9	97.6%	11
Fuel Cells	34	725	21	10	100.0%	10
Services Support	18	543	30	22	97.2%	3
Hydro	10	439	44	13	100.0%	3
Other	7	359	51	11	89.5%	3
Total/All SEOs	**462**	**30,711**	**66**	**15**	**97.0%**	**127**

The notion that a small number of (often mature and profitable) issuers account for large parts of proceeds is in line with DeAngelo, DeAngelo, and Stulz (2010) as well as the industrial population described by Fama and French (2008) (both for U.S. industrial firms), although the internationality and (emerging) industry nature of our sample seems to make its distribution particularly disparate. The ratio of primary shares is high, averaging 97% for the entire sample, and supports the view that capital demand is high. Firms in the United States, the United Kingdom and Australia are the most active issuers in the industry, accounting for 63% percent of SEOs, 40% of issue proceeds, and 47% of IPOs. In terms of sub-sector activity, wind and solar generated more than 60% of issue proceeds, while issuers in the field of efficiency recorded the most transactions (90).

Our analysis of the timing impact on the decision to conduct an SEO includes two parts: we first run logit regressions with market timing proxies used by Loughran and Ritter (1997) and Baker and Wurgler (2002). We further analyze key financial statement data surrounding the issue to detect a market-timing motive. If the latter is a major driver of the SEO decision, then the absence of an immediate cash need should show in managers' subsequent financing and investing decisions.

For our logit regressions, we pool all SEOs in firm-year observations and add up all proceeds of issuers with multiple issues per calendar year. We also require issuers to keep trading for a full calendar year after the issue year. This considerably shrinks our sample to 293 SEO-firm-years. Our dependent variable equals one if the company conducts one or more SEOs in any given year and zero if it does not.

We use the firm's market-to-book value (MTBV) and its prior and future market-adjusted return as market-timing indicators (independent variables). The MTBV is calculated at fiscal year-end prior to the calendar year in question and is truncated at 10. Prior returns are calculated up to the last day of the calendar year prior to the issue year. Similarly, future return calculations start with the first day of the calendar year following the issue year to obtain comparable, equally-sized firm-year observations. Both return metrics are adjusted by the return of the respective reference market index as suggested by Datastream, with no firm-specific risk adjustments. For firms listed less than the required period (e.g. 36 months) we calculate returns based on the close of the first trading day up to last day of the year prior to the issue year and treat it as the respective prior return (e.g. 36-month return). Firm-year observations with less than the required future return period (e.g. 36-month future returns for SEOs issued after 2007) are excluded from the respective analysis. As a proxy for the life-cycle stage of the company, we include the number of years the firm has been listed on the stock exchange.

As DeAngelo, DeAngelo, and Stulz (2010) note, the market-to-book value is an ambiguous market timing indicator because it may also capture "the existence and arrival of profitable investments at growth firms as argued by Jung, Kim, and Stulz (1996) and Carlson, Fisher, and Giammarino (2006)" (p.281). In the model used, however, we attribute the full life-cycle impact to the years listed variable and the market timing impact to the previously described market-to-book value and prior as well as future net of market returns. This ambiguity is mitigated by the fact that the market-to-book value plays a less prominent role in our logit results.

This figure illustrates our logit model variables using the example of Nordex SE. Grey-shaded areas mark years without SEO activity, which would cause the dependent variable to be zero. Company age relates to number of years the company has been listed on a stock exchange and is taken as of the calendar year-end prior to the year in question. Market-to-book values are measured at the last fiscal year-end immediately preceding the year in question. We use different return metrics (raw, market-adjusted) and time periods (12, 36 months). Share prices are rebased to 100 at 1 January 2004.

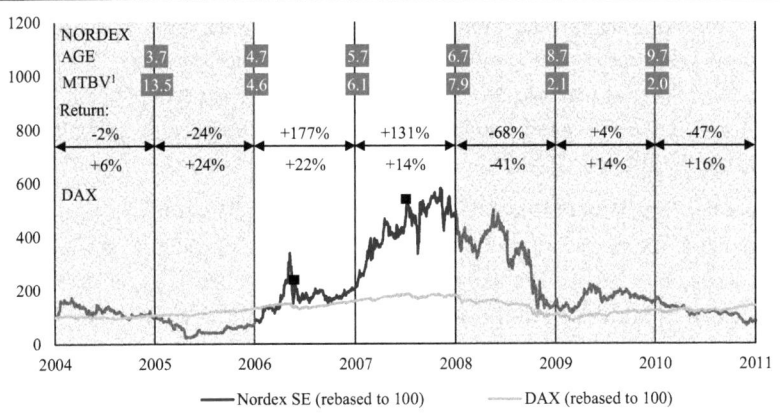

Source: Datastream, Thomson Reuters

[1] We consider all years the issuer has been listed on a stock exchange. We cut off the picture in 2004 due to a change in the fiscal year-end of Nordex SE (2003) and related illustrative purposes.

Figure 3.2: Logit regression variable illustration

Figure 3.2 graphically illustrates our approach using the example of wind turbine manufacturer Nordex SE. The company went public in March 2001 and conducted SEOs in May 2006 and July 2007 (two non-cash raising transactions took place in 2010). Nordex SE has experienced a rather volatile valuation history, with a maximum equity value of USD 2.4 billion in late 2007 and a current (equity) capitalization of USD 250 million (as of June 2012). For example, the model inputs for the first transaction would be a company age of 4.7 years, a MTBV of 4.6 and a 12-month net of market return of -48% (-24%-24%).

To test the statistical significance of our coefficients, we cluster standard errors by both time and issuer as suggested by Petersen (2009). We do so to account for time variation and differences in issuer characteristics. Market-wide shocks, such as the financial crisis, may cause our indicators to temporarily fluctuate on an economy-wide basis. Some business models may command consistently higher market-to-book ratios, and companies may choose differently as to when (company age) they go public.

3.2 Logit regression analysis

Table 3.2 reports our logit results partitioned in regressions based on 12-month (A.1-A.4) or 36-month (B.1-B.4) returns as well as with or without the life-cycle variable. We further provide regressions with a minimum market capitalization of USD 20m to check whether our results are driven by small issuers. For each regression, we report fitted coefficients together with t-statistics and marginal probabilities at sample means.

Our results document a positive relationship between the probability to conduct an SEO and the firm's market-to-book ratio. The statistical significance, however, is weak and fades away when introducing the years listed variable. This is possibly due to the previously described ambiguity of the market-to-book value as a timing indicator. Young growth firms are likely to command high relative valuations and thus the timing indicator may capture some of the effects previously recorded by the market-to-book ratio. The results are unchanged if we truncate at 15 instead of 10 or use calendar year-end values instead of fiscal year-end values (not reported).

Table 3.2: Logit regression results

Logit analysis of the seasoned equity offering decision in a given year as a function of the firm's previous fiscal year-end market-to-book value (MTBV), market-adjusted stock return over the prior and subsequent 12 (rows A.1 through A.4) or 36 months (rows B.1 through B.4), and the number of years listed. Rows A.1 through C.2 report estimated coefficients (with t-statistics and marginal probabilities evaluated at sample means) for logit models that pool all firm-year observations for the period 2000–2009. We use non-standardized, truncated (at 10) market-to-book values. Pseudo R-squareds reach a maximum of 0.061 for the model in row B.4. ***, **, and * denote significance levels at the 1%-, 5%-, and 10%-level, respectively.

		Intercept	MTBV	Prior stock return	Future stock return	Years listed	n=
A.1	M/B, adjusted 12-month returns						
	Coefficient	-1.06	0.05	-0.02	-0.52	-	817
	t-statistic	-3.73***	1.45	-1.55	-2.70***	-	
	Marginal prob.		0.01	0.00	-0.13	-	
A.2	M/B, raw 12-month returns						
	Coefficient	-1.04	0.05	-0.02	-0.50	-	817
	t-statistic	-3.76***	1.41	-0.99	-2.40**	-	
	Marginal prob.		0.01	-0.01	-0.13	-	
A.3	M/B, adjusted 12-month returns, years listed						
	Coefficient	-0.61	0.04	0.00	-0.49	-0.05	817
	t-statistic	-2.62***	1.36	-0.38	-2.69***	-2.48**	
	Marginal prob.		0.01	0.00	-0.13	-0.01	
A.4	M/B, adjusted 12-month returns, years listed, > USD 20m market cap.						
	Coefficient	-0.65	0.04	0.00	-0.41	-0.05	682
	t-statistic	-3.12***	1.05	0.02	-2.25**	-2.09**	
	Marginal prob.		0.01	0.00	-0.10	-0.01	
B.1	M/B, adjusted 36-month prior and future returns						
	Coefficient	-1.39	0.06	0.00	-0.35	-	546
	t-statistic	-4.36***	2.19**	-0.18	-3.74***	-	
	Marginal prob.		0.01	0.00	-0.07	-	
B.2	M/B, raw 36-month returns						
	Coefficient	-1.36	0.05	0.00	-0.42	-	546
	t-statistic	-4.95***	1.94*	0.12	-2.87***	-	
	Marginal prob.		0.01	0.00	-0.08	-	
B.3	M/B, adjusted 36-month prior and future returns, years listed						
	Coefficient	-0.93	0.04	0.00	-0.30	-0.05	546
	t-statistic	-3.10***	1.59	0.10	-4.22***	-2.80***	
	Marginal prob.		0.01	0.00	-0.06	-0.01	
B.4	M/B, adjusted 36-month returns, years listed, more than USD 20m market cap.						
	Coefficient	-0.82	0.03	0.01	-0.35	-0.06	456
	t-statistic	-2.89***	0.94	0.22	-3.11***	-4.15***	
	Marginal prob.		0.01	0.00	-0.07	-0.01	

C.1	Adjusted 12-month future return, years listed					
	Coefficient	-0.54		-0.43	-0.06	913
	t-statistic	-2.09**		-2.54**	-2.66***	
	Marginal prob.			-0.11	-0.01	
C.2	Adjusted 36-month future return, years listed					
	Coefficient	-0.86		-0.29	-0.06	629
	t-statistic	-3.02***		-4.85***	-3.31***	
	Marginal prob.			-0.05	-0.01	

We find no evidence for systematic stock price run-ups prior to the SEO. This was expected for our sample, but is remarkable in that it runs counter to the majority of cross-industry studies. However, it is in line with the notion that the primary motive of growth issuers is not market timing but capital need. The results for the prior stock return variable remain materially unchanged if we exclude any observation without a full stock return history (e.g. 8-month instead of 12-month history due to the firm's recent IPO).

By contrast, the future stock return is negatively correlated to the SEO probability and both statistically and economically significant. It remains qualitatively unchanged across all models (also when using other time periods, e.g. 24-months, or applying the variations described above). The economic significance, indicated by the marginal probability, is high and reaches a maximum of 13.5% in model A.1. We also find strong support for the negative relationship between the firm's age, proxied by the number of years listed, and the probability to conduct an SEO.

Due to the statistical insignificance of the market-to-book ratio and the prior excess return, we re-run the regression with only the future stock return and the years listed variable (rows C.1 and C.2). Again, the results are materially unchanged and document a strong influence of both factors on the decision to conduct an SEO. The marginal probabilities, however, differ between the 12-month and 36-month return variable, similar to the broader regressions. We thus use both models to provide a sense of the economic impact of both variables on the decision to conduct an SEO.

Table 3.3: Estimated SEO probabilities

Estimated probability to conduct an SEO depending on hypothesized values for the future excess return and number of years listed. Values are based on models C.1 and C.2 of Table 0.7 and include the net-of-market returns calculated over a 12-month (C.1) or 36-month (C.2) period following the issuing calendar year. The years-listed variable is the truncated (20) and based on the number of years the company has been listed on the stock exchange.

12-month future return, years listed (model C.1)						36-month future return, years listed (model C.2)					
Excess stock return	Number of years listed					Excess stock return	Number of years listed				
	1	5	10	15	20		1	5	10	15	20
-75%	43.3	37.7	31.3	25.5	20.4	-75%	33.1	28.0	22.5	17.7	13.8
-50%	40.7	35.3	29.0	23.5	18.7	-50%	31.5	26.6	21.2	16.7	12.9
-25%	38.1	32.9	26.9	21.6	17.2	-25%	29.9	25.2	20.0	15.7	12.1
0	35.6	30.5	24.8	19.9	15.7	0	28.4	23.9	18.9	14.8	11.4
25%	33.2	28.3	22.9	18.2	14.3	25%	27.0	22.6	17.8	13.9	10.7
50%	30.9	26.2	21.0	16.7	13.1	50%	25.6	21.3	16.8	13.0	10.0
75%	28.6	24.2	19.3	15.2	11.9	75%	24.2	20.1	15.8	12.2	9.4

Table 3.3 provides estimated probabilities based on model C.1 and C.2 of Table 3.2 with various hypothesized values for excess future stock returns and the number of years listed. A firm listed for one year with zero excess future return (no timing opportunity) has a probability of 35.6% to conduct an SEO (28.4% for the 36-month model). For firms listed for 20 years, the probability falls to 15.7% (11.4%). Young companies with significant timing opportunities (future excess return of -75%) have a high probability of 43.3% (33.1%) to issue equity, whereas a mature firm listed for 20 years with poor timing opportunities (excess return of +75%) has a probability of just 11.9% (9.4%). Thus, both future returns and the life-cycle stage have a strong impact on the probability to conduct an SEO.

Summarizing this subsection, we conclude that the results are predominantly in line with our hypothesis. Equity does not tend to be issued after periods of excessive share price growth or particularly high valuations. Young companies more often tap capital markets than mature ones, presumably due to stronger growth prospects. However, the consistently negative returns following the offering could be read as management's anticipation of challenging prospects and superior information. An alternative explanation is associated with changes in the risk profile of the issuer, which we will investigate next.

3.3 Risk dynamics

Firm-level studies of market timing have been controversially discussed, particularly with regards to their consideration of risk. Fama (1998) argues that most firm-level returns prior and subsequent to the event are sensitive to the methodology and benchmark used. Jegadeesh (2000), on the other hand, shows that negative post SEO performance is robust to a variety of benchmarks. Eckbo et al. (2000) argue that the post-issue underperformance is mainly due to lower systematic risk caused by lower leverage and increased stock liquidity. Similarly, but in an option pricing framework, Carlson, Fisher, and Giammarino (2006, 2010) posit that stock price patterns surrounding SEOs are due to increasing risk prior to the issue (as the growth option moves into the money) and declining risk (as the growth option is exercised) after the issue.

Our research setup mitigates some of these short-term risk effects by using calendar years surrounding the issue. In particular, stock price run-ups and increasing risk immediately prior to the issue should be less of a problem. Our results support this view as we do not find significant stock price run-ups. However, our significantly negative post-issue stock returns could be the result of reduced systematic risk after the issue. To check if our results are driven by a risk change, we calculate betas for the calendar year prior and subsequent to the issue year. We drop any issue with a confounding event in the prior or subsequent year and without a full year's stock price history. This reduces the sample to 129 issues. Panel A from Table 3.4 shows the results from the standard market model, which corresponds to our indicator in the logit regressions. Panel B reports betas based on a Fama-French three-factor model with factors (HML, SML) derived from international mimicking portfolios (Russell) as in Faff (2003). In panel C we replace the individual market index with a single global index for all issues (Russell Global Index). We test for normal distribution with a Shapiro-Wilk test (Shapiro and Wilk, 1965) and medians with a Wilcoxon test (Wilcoxon, 1945).

We do not find any indication for a significant risk change. Beta levels and changes are similar across all three models and remain statistically insignificant. Therefore, we do not believe that our negative post issue performance is driven by risk dynamics. Our previous results remain unchanged. We will now turn to use-of-funds analysis to find additional indicators on the presence (or absence) of market timing.

Table 3.4: Risk changes surrounding SEOs

This table reports beta calculations before and after the issue of seasoned equity offerings. We include only issues without confounding events in the calendar year before and after the issue. We estimate betas over a 12-month period (250 trading days) in the calendar year before and after the issue year. We estimate three models: panel A shows the results of a market model regression using the respective market index of the issuer, as suggested by Datastream. Panel B displays results of a Fama-French three-factor regression, with the respective market index and Russell style portfolios as benchmark. We calculate factors as in Faff (2003). Panel C replaces individual market indices with a single global market index (Russell Global Index). Changes in means are tested with a standard t-test, medians with a non-parametric Wilcoxon test. Means are tested for normal distribution using the Shapiro-Wilk-test. ***, **, and * denote significance levels at the 1%-, 5%-, and 10%-level, respectively.

		Beta before issue	Beta after issue	Beta change	p-value
Panel A: Market-model with respective market index as benchmark (n=129)					
Equity beta	Mean	0.874	0.881	0.007	(0.913)
	Median	0.889	0.888	-0.001	(0.703)
Panel B: Fama-French three-factor model with respective market index as benchmark (n=129)					
Equity beta	Mean	0.914	0.906	-0.008	(0.887)
	Median	0.860	0.914	0.054	(0.948)
Panel C: Fama-French three-factor model with single global index as benchmark (n=129)					
Equity beta	Mean	0.928	0.905	-0.023	(0.737)
	Median	0.890	0.815	-0.074	(0.792)

3.4 Use-of-funds analysis

If timing is a major driver of the SEO decision, then the lack of an imminent financing need should show in management's subsequent investing and/or financing decisions. Specifically and in particular, elevated cash balances should point to market timing, all other things equal. To mask market timing and diffuse shareholder concerns regarding the inefficient stockpiling of cash, management could even engage in negative net present value projects to increase assets under management and power (Jensen, 1986). As controlling for negative net present value projects is virtually impossible, we analyze managements' financing decisions surrounding the SEO. Additional debt financing is indicative of a cash need rather than market timing, as interest payments would lower free cash flow available to management in case of a negative net present value project. Rather, more debt points to increased borrowing capacity after the receipt of SEO proceeds, and the pursuit of growth opportunities that were previously non-executable due to financing constraints. The "price" to management by means of increased interest payments and reduced free cash flow makes it unlikely that such projects are purely motivated by discretionary spending.

Table 3.5 summarizes our results with respect to SEO proceeds and cash balances. We keep the clustering of within-year SEOs and analyze a sample of 293 firm-year SEOs. Due to the internationality of our sample and the corresponding currency effects, we convert issuer proceeds into the respective currency as of date of the issue. All other data is retrieved in local currency terms. We provide medians on various ratios for two groups: the first contains all firms, for which data for the respective indicator is available; the second contains only those firms, for which data on all indicators are available.

The number of observations in Table 3.5 shows that data availability is an issue: full information (cash, total assets, CapEx, total debt, dividends) for the three years surrounding the SEO is available only for half of the sample (157 issues). This is partly due to the fact that some firms issue within their IPO year and prospectus data is not always immediately brought forward to the database. The large number of small firms with limited and/or infrequent reporting aggravate the problem. However, the distributions of most indicators are similar and we will go into more detail in the few cases where they diverge. In the following, we will use the larger sample for our discussion of the results and selectively report the corresponding value of the trimmed sample in brackets.

The analysis shows that the median SEO size over the decade is 19% of total assets. Cash levels are at 10% of total assets before the issue year, increase to 13%–14% in the SEO year and decrease in the two consecutive years to pre-SEO levels. All other things equal, approximately 25% of SEO proceeds are held as extra cash in the year of the SEO. This, however, ignores the idea that larger firms typically require larger cash reserves to operate as suggested by Opler et al. (1999). We therefore calculate cash balances adjusted for asset growth; we multiply the cash balance before the SEO with the asset growth over the respective period and deduct the respective cash balance after the SEO. Any difference should be surplus cash not accounted for by asset growth. We then divide the absolute difference by the SEO proceeds to isolate the relative cash increase.

Table 3.5   Impact of SEO proceeds on issuer cash balances

This table provides actual and pro-forma values of the cash-to-total assets (TA) ratio for renewable energy firms conducting an SEO over the period 2000–2009. The numbers in curly brackets indicate the year relative to the SEO year [0]. The raw change in cash of rows 8–10 is the absolute difference between cash positions of the respective years. For rows 11–13, we multiply the pre-SEO cash level with the asset growth over the respective period, deduct it from the cash level of the respective period after the SEO and subsequently divide it by the SEO proceeds. Rows 14–18 show ratios of cash to total assets if the company had not received the SEO proceeds and left all other decisions unchanged. The fixed issue number is 157.

		All issues with data available	(n=)	Fixed issue number
1.	Median cash [-1]/TA [-1]	10.1%	(204)	9.4%
2.	Median cash [0]/TA [0]	13.4%	(207)	13.5%
3.	Median cash [+1]/TA [+1]	12.2%	(183)	11.4%
4.	Median cash [+2]/TA [+2]	10.8%	(140)	-
5.	Median proceeds/TA [-1]	27.4%	(281)	26.7%
6.	Median proceeds/TA [0]	19.1%	(285)	17.4%
7.	Median proceeds/TA [+1]	17.1%	(259)	15.5%
8.	Median raw change in cash [-1;0]/proceeds	23.7%	(192)	25.2%
9.	Median raw change in cash [-1;+1]/proceeds	20.7%	(168)	24.6%
10.	Median raw change in cash [-1;+2]/proceeds	20.3%	(128)	-
11.	Median abnormal change in cash [-1;0]/proceeds	1.0%	(192)	1.8%
12.	Median abnormal change in cash [-1;+1]/proceeds	0.0%	(168)	0.6%
13.	Median abnormal change in cash [-1;+2]/proceeds	-15.7%	(128)	-
	No SEO proceeds			
14.	Median pro-forma cash [+1]/TA [+1]	0.0%	(183)	-1.0%
15.	Percent with pro-forma cash [+1]/TA [+1] < 0	54.6%	(183)	54.1%
16.	Percent with pro-forma cash [+1]/TA [+1] < 5%	66.1%	(183)	66.2%
17.	Percent with pro-forma cash [+1]/TA [+1] > 20%	14.8%	(182)	13.4%
18.	Percent with pro-forma cash [+1]/TA [+1] > 30%	10.9%	(182)	10.2%

The results in rows 11–13 show that the net-of-asset growth in cash is marginal, for the median issuer between 0% and 2%. Moreover, the ratio turns strongly negative in the second year after the SEO, indicating that cash balances actually shrink in absolute terms in the second year after the SEO for the majority of issuers. The notion that cash level increases are small and transient in nature is also supported by the pro-forma cash balances calculated in rows 14–18. They show hypothesized values for cash to total assets if companies had not received the SEO proceeds and left all other decisions unchanged. More than half of the sample would have run out of cash and approximately two thirds would have relative cash holdings below the pre-SEO level. However, to the right of the distribution, there is a nontrivial amount of issuers with large cash levels: approximately 14% of our sample issuers maintain net-of-asset-growth cash-to-assets ratios above 20% and approximately 10% have ratios above 30%.

Our preliminary conclusion is that the raw increase in cash, observed in the SEO year, is mainly due to an increased asset base, which demands more cash resources to operate. Accounting for the asset increase, extra cash holdings are marginal for the median issuer. This lends additional support to the idea that the primary motive of the issuing management is not market timing and that the lack of an imminent financing need does not lead to the stockpiling of cash.

Table 3.6 provides evidence on the most important use of funds of the sample: capital expenditures. Our median issuer steps up CapEx in the issue year and continues to invest the year after the SEO. Over the three-year period, investments add up to 90% of issue proceeds of the total sample (with data available for respective ratio) and 146% of proceeds of the restricted sample (data available for all ratios).

Incremental CapEx, however, calculated as the raw change in CapEx divided by SEO proceeds, is relatively low at 5% to 10% of issue proceeds in the issue and subsequent year. This is the only indicator where the trimmed sample provides significantly different results than our larger sample. We therefore also provide the distribution of CapEx for the three years surrounding the issue relative to SEO proceeds in rows 4–10. The distribution is u-shaped and shows that 30% invest less than half, while 40% spend more than 200% of issue proceeds in the three-year period. Finally, we also calculate pro-forma cash balances in case the firm had not received the proceeds and not increased CapEx after the issue. If we revers these cash flows, cash balances would shrink to levels below those of the pre-SEO year (7%).

Note, however, that he CapEx indicator is difficult to interpret as investments may come in form of large, binary capital outlays. Some firms may make investments that increase in size throughout the project and require the firm to raise additional capital to continue investing or replenish liquidity. Others may issue early and undertake a large investment only in the first or second year after the issue. As such, using CapEx levels prior to the issue as a base for recurring levels of CapEx can be misleading. If this is the case, then the ratios in the pre-SEO year would not represent unbiased benchmarks.

Table 3.6: Impact of SEO proceeds on issuer capital expenditures

This table provides actual and pro-forma values of the cash-to-total assets ratio for renewable energy firms conducting an SEO over the period 2000–2009.The numbers in curly brackets indicate the year relative to the SEO year [0]. Rows 1-9 use absolute values of CapEx, rows 10–14 use changes in CapEx. Rows 15 and 16 show cash-to-total assets ratios, assuming that the company had not received the SEO proceeds and allocated any CapEx increases to the cash position. The fixed issue number is 157.

		All issues with data available	(n=)	Fixed issue number
1.	Median CapEx [-1]/proceeds	25.2%	(282)	29.1%
2.	Median CapEx [0]/proceeds	33.5%	(285)	46.2%
3.	Median CapEx [+1]/proceeds	31.6%	(256)	70.3%
	Total CapEx [-1;+1]/proceeds distribution			
4.	0%< CapEx/proceeds <25%	17.5%	(252)	14.0%
5.	25%< CapEx/proceeds <50%	12.7%	(252)	12.1%
6.	50%< CapEx/proceeds <75%	7.9%	(252)	5.7%
7.	75%< CapEx/proceeds <100%	6.0%	(252)	5.7%
8.	100%< CapEx/proceeds <200%	16.7%	(252)	16.6%
9.	CapEx/proceeds >200%	39.3%	(252)	45.9%
10.	Median absolute CapEx change [-1;0]/proceeds	4.9%	(282)	7.3%
11.	Median absolute CapEx change [0;+1]/proceeds	0.4%	(274)	2.8%
12.	Percent of Issuers with CapEx [-1;0] decrease	36.5%	(282)	32.5%
13.	Percent of Issuers with CapEx [-1;0] increase up to 100%	52.5%	(282)	55.4%
14.	Percent of Issuers with CapEx [-1;0] increase > 100%	11.0%	(282)	12.1%
	No SEO proceeds, no CapEx increase			
15.	Median pro-forma cash [+1]/TA [+1]	7.0%	(180)	7.3%
16.	Percent with pro-forma cash [+1]/TA [+1]<0	29.4%	(180)	29.3%

High CapEx spending does not rule out market timing. Additional funds could simply have been used on projects that otherwise would not have been undertaken. In combination with additional debt and its disciplinary effect on management, however, CapEx could signal a previous financing constraint that is mitigated with additional equity via the offering.

Table 3.7 provides key ratios on the sample's financing decisions over the three-year period. Our median issuer markedly steps up debt levels and almost doubles its debt-to-total assets ratio in the issue year. Firms also continue to lever up in the second year with relative debt level increases of similar size. However, there is again a nontrivial share of the sample that strongly reduces debt. Between 14% and 15% of the sample reduce debt by more than 50%.

Table 3.7: Impact of SEO proceeds on issuer capital structure

This table provides actual and pro-forma values of the cash-to-total assets ratio for renewable energy firms conducting an SEO over the period 2000–2009. The numbers in curly brackets indicate the year relative to the SEO year [0]. Rows 15 and 16 show cash-to-total asset values if the company had not received the SEO proceeds and any debt increases were not available to the firm. In rows 17 and 18, all dividend payments in the year of or after the SEO are added back to the cash position. The fixed issue number is 157.

		All issues with data available	(n=)	Fixed issue number
1.	Median total debt [-1]/TA [-1]	4.9%	(282)	6.1%
2.	Median total debt [0]/TA [0]	9.5%	(285)	10.1%
3.	Median total debt [+1]/TA [+1]	13.1%	(258)	15.9%
4.	Percent of issuers with debt increase [-1;+1]	51.0%	(255)	56.7%
5.	Percent with no change in debt [-1;+1]	12.5%	(255)	10.8%
6.	Percent with debt decrease less than 25% of proceeds	19.2%	(255)	15.3%
7.	Percent with debt decrease between 25% and 50% of proceeds	3.1%	(255)	2.5%
8.	Percent with debt decrease > 50%	14.1%	(255)	14.6%
	No SEO proceeds, no increase in debt			
9.	Median pro-forma cash [+1]/TA [+1]	-7.5%	(183)	-7.8%
10.	Percent with pro-forma cash [+1]/TA [+1] < 0	59.6%	(183)	59.9%
11.	Percent of dividend payers [-1]	17.5%	(263)	16.6%
12.	Percent of dividend payers [0]	14.6%	(268)	15.3%
13.	Percent of dividend payers [+1]	16.0%	(244)	18.8%
14.	Median dividend [-1]/TA [-1] of dividend payers	1.5%	(46)	0.6%
15.	Median dividend [0]/TA [0] of dividend payers	1.0%	(39)	0.7%
16.	Median dividend [+1]/TA [+1] of dividend payers	0.9%	(39)	0.3%
	No SEO proceeds, no dividends			
17.	Median pro-forma cash [+1]/TA [+1]	0.9%	(173)	0.8%
18.	Percent with pro-forma cash [+1]/TA [+1]<0	46.8%	(173)	46.8%

When calculating the pro-forma cash balances, we see the strong effect of debt on cash: the majority of issuers would have strongly negative cash-to-total assets balances if no additional debt was raised. In this ratio, we deduct both the SEO proceeds and any debt increases in the issue and the subsequent year, assuming that additional equity was necessary to acquire the incremental debt financing.

An alternative means of financing would be to withhold any dividend payments to fund new projects. If positive net present values are imminent, it should be more valuable to invest those funds than to redeem them to shareholders. Both the percentage of dividend payers and their respective dividend payment are low: only 15% of issuers pay a dividend in the issue year in the magnitude of approximately 1% of assets. Withholding dividends does reduce the shortfall in cash as indicated by the pro-forma cash calculation in row 18, but the majority of issuers would still maintain below pre-SEO cash balances.

The results of the use-of-funds analysis argue against a widespread market timing motive. We interpret the indicator as supportive of the notion that stockpiling of cash is the exception rather than the rule. Cash level increases in the year of the issue are small and quickly revert to prior levels. Without the proceeds, more than half the issuers would have run out of cash and two thirds would have significantly reduced cash levels (when considering asset growth). However, there is a minority of issuers that retain large parts of SEO proceeds in their cash balances, which could indicate the lack of an imminent financing need.

We believe the primary use of those funds is capital expenditure. While a difficult indicator to interpret, 40% of the issuers in our sample invest twice the SEO proceeds over the three years. Also, when holding back any additional CapEx after the issue, the shortfall in cash would be largely compensated for. More than half of the issuers step up debt levels after the issue and if they do, they tend to do so considerably. Debt levels almost double for our median issuers. Dividends tend to play a minor role.

3.5 Conclusion

We analyze the impact of market timing on the decision to conduct an SEO in a particular growth industry. We run logit regressions on a set of 462 renewable energy SEOs over the period 2000–2009 and analyze key financial statement data surrounding the SEO to detect market timing. Our analysis is based on the notion that both market timing and the need for growth capital are key motives for the decision to conduct an SEO, but that their relative importance may change throughout the life-cycle of an industry.

We find no support for stock price run-ups or elevated market-to-book ratios prior to the issue. This was expected but runs counter to most cross-industry, firm level studies. It is in line with the idea that the primary motive for firms in a growth setting is capital demand. In addition, we find strong support for the life-cycle indicator, which posits that young firms are more likely to issue equity than mature firms. However, we find strongly negative excess returns after the issue. The effect is not driven by a risk change and is robust to various time periods and model specifications. Its impact is substantial: according to our model, a renewable energy company listed for one year with a negative future return of -75% has a probability of 33% to conduct an SEO, versus 9% probability for a company listed for 20 years with a subsequent return of +75%.

Our use-of-funds analysis shows that, without the SEO proceeds, most issuers would have run out of cash in the year following the SEO. Most of the incremental cash held by the firm corresponds to an increase in asset size, and slightly elevated cash balances are transient in nature. We find that most issuers spend quickly and heavily on capital expenditures; almost 40% spend more than twice of the issue proceeds in the three years surrounding the SEO. We also find that our median issuer markedly steps up debt financing (relative to assets) after the issue. All this supports the notion that the stockpiling of cash is the exception rather than the rule. We also believe that the increasing use of debt is an argument against discretionary spending of the issuing firm's management and the systematic timing of the market. However, there is a non-trivial minority of firms with substantially elevated cash balances after the SEO, limited CapEx spending and a substantial pay down in debt.

Our results of the use-of-funds analysis are only partly in line with those recorded by DeAngelo, DeAngelo, and Stulz (2010). We find similar values for cash levels, SEO size and pro-forma values of cash if the SEO proceeds are removed. Investing and financing, however, show a different behavior. The number of issuers in our study with CapEx and debt increases over the three year period is smaller, but those who do expand tend to do so more aggressively. We believe capital need is the primary motive, either to significantly expand operations or to survive an initial life-cycle period where operating cash flow is low and more successful business models are separated from poorer ones. This environment may offer timing opportunities for some issuers.

4 Signaling with Convertible Debt in the Renewable Energy Industry

Convertible debt is an important component of the range of financing options available to corporate managers. The setup of straight debt with an attached call option reduces the cost of debt when traded as a bond, while strengthening the equity capital base when converted into equity. Convertible debt can be a valuable financing tool when the risk level of the issuer is difficult to estimate. It can act as a hedge for investors, because changes in the value of the debt component are (at least partially) offset by value changes in the option (Brennan and Schwartz, 1988). Moreover, convertibles can be structured quite flexibly to match different cash flows and efficiently finance sequential investment projects (Mayers, 1998).

As illustrated by their frequent use in the venture capital industry, such features make hybrid securities particularly attractive as a financing structure for high-risk companies (Kim, 1990; Schmidt, 2003; Trester, 1998).[28] They can play an important risk-mitigating role in the development of new technologies in emerging industries. However, due to the private nature of the venture capital industry, research on the structure, signaling effect, and valuation impact of convertible debt in such high-risk industries has thus far been limited. Despite the importance of these types of companies for economic development (Lerner, 2009), as well as academic and political debate over their funding (Cressy, 2002; OECD, 2006), most research has concentrated primarily on cross-industry studies focused geographically on the U.S.

In this chapter, we take a life-cycle and industry perspective to convertible debt, and do so for the following reasons. Lewis, Rogalski, and Seward (2003, p. 155) argue that "Because convertible debt can be structured to mitigate several different combinations of debt- and equity-related costs of external finance, an empirical examination of average valuation effects for the full universe of issuers is likely to be uninformative." We believe these differences are driven substantially by life-cycle effects. Furthermore, we hypothesize that emerging industry issuers have a stronger focus on signaling quality when adverse selection costs are high; they are also more likely to put a strong emphasis on cash conservation,

---

28 The use of hybrid securities (such as, e.g., convertible preferred equity) has been linked to the success of the U.S. venture capital industry (Black and Gilson, 1998).

risk mitigation, and access to finance than more mature companies. All of these elements have security design and valuation implications.

Furthermore, emerging industries play a key role in innovation and economic growth. Some research has argued that young and innovative companies tend to be priced out of the capital markets or face prohibitively high financing costs due to their risk characteristics and levels of asymmetric information (Cressy, 2002). Convertible debt can alleviate some of these problems, and could thus improve financing for an important group of companies.

Our analysis focuses on renewable energy issuers. This industry, with its distinct growth and risk profile, provides unique insights into how emerging industry issuers use convertible debt securities. The potentially large energy market and the desire for a more sustainable means of energy production create significant growth opportunities. At the same time, market volatility has been fueled by concerns about competition from conventional energy carriers,[29] uncertainty about the prospects of various renewable energy technologies, and market distortions caused by government intervention. Government subsidies and the capital intensity of the industry have led to issue sizes that justify the substantial transaction costs (structuring, pricing, distribution) associated with convertible debt issuance. Such costs may be a deterrent to smaller issuers. The industry's capital intensity also provides collateral to the often secured debt part of the convertible, which may make these vehicles more attractive here than they would be in other industries.

Finally, the industry is currently at a pivotal life-cycle stage. As governments begin to withdraw their financial support, convertibles may be particularly suitable instruments to mitigate the risks associated with this process.

This study contributes to three primary strands of literature. First, we add to the research on the motives for issuing convertible debt. In line with the proponents of sweetened debt (Brennan and Kraus, 1987; Brennan and Schwartz, 1988; Green, 1984; Mayers, 1998) and delayed equity (Stein, 1992) models, we offer a complimentary life-cycle perspective for issuing convertible debt. We further add to the signaling research on convertible security design as suggested by Kim (1990), Davidson, Glascock, and Schwarz (1995), and Jung and Sullivan (2009).

Second, we contribute to event study research on stock returns around convertible debt issues. Based on the signaling model of Myers and Majluf (1984), the

---

29 Examples are natural gas in the U.S. and nuclear energy in France and Japan.

default interpretation of empirical results is that equity offerings are perceived as stronger signals of firm overvaluations than straight debt offerings, and are accompanied by more negative announcement returns. Convertible issues rank somewhere in between depending on security design (Dann and Mikkelson, 1984; Lewis, Rogalski, and Seward, 1999, 2003; Mikkelson and Partch, 1986).

Third, we contribute to the literature on finance and growth. Rajan and Zingales (1998) show that financial development has an impact on industrial growth. Levine (2005, p. 922) conducts a literature review of the relationship between finance and growth, and concludes that "technology innovation, for example, may only foster growth in the presence of a financial system that can evolve effectively to help the economy exploit these new technologies." We provide new insights into risk-mitigating securities that should more effectively mirror the risk and return distributions of emerging industry issuers.

Our sample consists of 44 convertible debt and 285 seasoned equity offerings over the 2001–2010 period. Our results show that convertibles are structured to resemble debt rather than equity. With a low conversion probability of 19%, their ranking is well below that found in prior cross-industry studies. At the same time, they incur strongly negative announcement returns of -3.8% over the three-day period surrounding the announcement, which exceeds the negative returns exhibited by seasoned equity issuers (-2.3%).

We also find that convertible issues are preceded by stock price declines. They exhibit higher volatility and financial distress-related indicators (leverage, Altman z-scores) than seasoned equity issuers. In line with the emerging nature of the industry, profitability and cash flow generation is low: almost half of all issuers analyzed exhibited negative operating earnings and cash flow prior to the issue.

We interpret our results as a failure by convertible debt issuers to credibly signal firm quality to the market. High adverse selection costs (volatility, stock price decline) seem to induce issuers to use debt-like securities to provide signals about firm quality. These signals tend to fail, however, because they are perceived as indicative of issuers having been priced out of equity markets due to excessively high adverse selection costs. This notion is further supported by higher financial distress costs, which should, if all other things remain equal, warrant a more equity-like financing instrument.

4.1 Literature review
4.1.1 Motivations for issuing convertible debt

Green (1984) argues that convertibles can help mitigate investment choice conflicts between bond- and stockholders. Because equity can be viewed as a call option with a strike price equal to the face value of debt, and option values typically increase with volatility, stockholders have an incentive to invest in particularly risky, even negative net present value projects at the expense of debtholders. This misalignment of incentives can be ameliorated by having bondholders share some of the upside potential (via the convertibility option) that can result from risky projects.

Brennan and Kraus (1987) and Brennan and Schwartz (1988) also show that convertibles can reduce asymmetric information costs between companies and outside bond investors. If there is a high level of uncertainty about an issuing company's true risk level investors will typically require a premium to compensate for that additional risk. Convertibles can provide a hedge against this problem by setting up the bond and call options to respond differently to changes in risk. For example, while the bond value would decrease, the embedded call option would increase as the risk level increased.

Stein (1992) argues that convertibles are issued by companies with equity financing needs who view the asymmetric information costs from issuing straight equity as prohibitively high. By issuing convertibles and forcing conversion once the stock price is at or beyond the strike price, managers can reduce the costs associated with equity issues. The model thus relies heavily on the assumption that convertibles are callable to lock-in funds once the stock price allows.

Mayers (1998) demonstrates that using convertibles for funding risky projects can be more favorable than using either short- or long-term debt. A short-term debt series incurs increased transaction costs, while a long-term bond lacks an abandonment option, which can be vital in the case of an unprofitable project. However, this again requires convertibles to be callable by the issuing firm in order to keep funds in the company.

Lewis, Rogalski, and Seward (1999) argue that the diverging models are not mutually exclusive, but rather depend on a firm's relative financing costs. According to this proposition, companies facing high equity-related costs are more likely to issue instruments resembling debt, while companies facing high debt financing costs are more likely to choose equity-linked instruments.

Empirical tests of these motives have found mixed results. Dutordoir and Van de Gucht (2009) provide an overview of research in this field. They conclude that earlier survey-based evidence mainly supports the delayed equity viewpoint (Stein, 1992), while more recent surveys are consistent with both the delayed equity and the sweetened debt hypotheses (Brennan and Kraus, 1987; Brennan and Schwartz, 1988; Green, 1984; Mayers, 1998). Quantitative analyses consist mainly of two methodologies: (i) event studies that link stock price reactions to the structure of the convertible (Abhyankar and Dunning, 1999; Ammann, Fehr, and Seiz, 2006; Wolfe, 1999) and (ii) multivariate regressions on the debt or equity features of the convertibles at issue, calculated based on Black and Scholes (1973) option pricing methodology (Dutordoir and Van de Gucht, 2009; Lewis, Rogalski, and Seward, 1999). Results were, again, mixed, with support for both the sweetened debt and the delayed equity hypotheses.[30]

4.1.2 Announcement returns and signaling

A growing body of research documents negative market reactions to convertible debt issue announcements. De Roon and Veld's (1998) literature review finds negative returns ranging from -0.6% to -2.3% for a two-day period around announcements.[31] More recent evidence has found even stronger negative returns, which have been linked to the rise of hedge funds (Duca et al., 2012). Relative to other financing events, convertible debt typically results in less negative returns than equity issues, but more negative returns than straight debt issues (Dann and Mikkelson, 1984; Eckbo, 1986; Lewis, Rogalski, and Seward, 1999; Mikkelson and Partch, 1986). This is in line with Myers and Majluf (1984) signaling model, where equity-like security offerings are perceived by the market as stronger signals of firm overvaluations.

Moreover, Davidson, Glascock, and Schwarz (1995) and Lewis, Rogalski, and Seward (2003) show that these valuation effects hold even across convertible

---

30 Dutordoir and Van de Gucht (2009) provide evidence of regional preferences for the design of convertibles. They show that European convertibles are predominantly structured to resemble debt.
31 Note that this includes statistically significant returns only. As exceptions, Kang et al. (1995) and Kang and Stulz (1996) find significantly positive announcement returns for Japanese convertible issues.

debt issues, i.e., more equity-like structured convertibles trigger more negative returns. Kim (1990) argues that companies with favorable earnings expectations are likely to issue convertibles with low conversion ratios to benefit from the future earnings. Firms with low conversion ratios, on the other hand, aim to share risk, and thus structure their securities in a more equity-like fashion. Davidson, Glascock, and Schwarz (1995) and Jung and Sullivan (2009) extend Kim's model by a growth component, and show that capital markets do respond to the signal conveyed by a security's design. They find that firms with higher growth expectations experience less negative valuation effects than firms with lower growth prospects.

4.2 Data and methodology

Given the lack of an industry code in standard industrial classification systems, we base our analysis on a dataset from Bloomberg New Energy Finance (BNEF), and from supplementary constituents of the Wilderhill New Energy Innovation industry index, which is also administered by Bloomberg. We obtain equity and convertible debt transaction data for the 2001-2010 period, including convertible bonds, privately placed notes, and convertible preferred equity, from the Thomson Reuters database. Preferred equity is typically structured more like debt in the U.S. than in Europe, and is frequently classified as debt.

We only consider issuers who have been listed on a stock exchange for at least 210 trading days, and have displayed no confounding equity or equity-linked events during this period. For liquidity reasons, we exclude issues below USD 5 million. To calculate option deltas, we further impose data availability requirements pertaining to financial statement data and to transaction data for convertible debt. We ultimately obtain a sample of 44 convertible debt and 285 seasoned equity offerings.

Table 4.1 provides an overview by year of issue and issuer domicile. Convertible issues are clustered in the 2007-2009 period. We note that nearly 60% of convertible transactions occurred in the U.S., while less than 30% of seasoned equity offerings were conducted there.

Table 4.1: Descriptive convertible debt sample characteristics

This table provides descriptive statistics for our sample of 44 convertible debt and 285 seasoned equity issues from 2001–2010. We obtain issuer names from Bloomberg New Energy Finance (BNEF), transaction information from Thomson Reuters, and financial statement and market price information from Datastream. We only consider issuers that have been trading for at least 210 trading days, and who have not experienced any equity or equity-linked financing events during this period.

Year	CDO[1] No.	CDO[1] Vol.	SEO No.	SEO Vol.	Country	CDO[1] No.	CDO[1] Vol.	SEO No.	SEO Vol.
2001	-	-	1	211	U.S.	26	3,619	90	7,477
2002	1	12	5	310	China	7	2,408	20	10,348
2003	1	29	8	868	Germany	2	868	28	3,022
2004	1	100	10	937	Canada	2	142	15	660
2005	1	6	14	3,133	Taiwan	2	42	13	2,374
2006	5	388	29	3,810	Spain	1	280	7	6,250
2007	11	3,459	63	12,802	France	1	271	6	1,248
2008	11	2,089	54	5,163	India	1	150	1	179
2009	8	1,238	59	8,984	Japan	1	32	6	1,212
2010	5	515	42	7,094	Other	1	24	99	10,542
Total	**44**	**7,836**	**285**	**43,312**	**Total**	**44**	**7,836**	**285**	**43,312**

[1] Includes twelve convertible bonds, twenty-four convertible notes/debentures, and eight convertible preferred securities.

4.3 Industry convertible debt structure

Our first step is to calculate the conversion probability of the issues. Multiple methodologies have been used for this purpose. For example, Beatty and Johnson (1985) analyze the potential to force conversion for callable convertible bonds by calculating call price divided by conversion value. Kuhlman and Radcliffe (1992) measure the motivation of bondholders to convert bonds by calculating conversion price divided by stock price. Moreover, Davidson, Glascock, and Schwarz (1995) estimate time to conversion based on stock growth expectations, and argue for a negative correlation between length of time and how debt-like the issue is.

Jung and Sullivan (2009) propose a different proxy for growth expectations based on expected conversion time, yield advantage, and conversion premium. They posit a positive correlation between growth expectations and the level of debt-likeness. The most frequently used method is to calculate the option's delta value, i.e., the sensitivity of the option's value to the underlying stock price, which is based on Black and Scholes (1973):

$$\Delta = N\left(\frac{\ln\left(\frac{S}{B}\right) + \left(r - \delta - \frac{\sigma^2}{2}\right)T}{\sigma\sqrt{T}}\right) \quad (1)$$

where N(•) is the cumulative probability under a standard normal distribution, S is the underlying stock price, B is the conversion price, r is the continuously compounded interest rate, δ is the continuously compounded dividend yield, σ is the underlying's stock volatility (standard deviation), and T is the convertible debt's maturity in years measured at the issue date. For our international sample, we approximate r with a five-year government bond of the respective issuer country, and we use the issuer's dividend yield at the fiscal year-end prior to the issue as a proxy for the expected future dividend yield.[32]

Figure 4.1 shows the sample's delta distribution. The average (median) conversion probability is 19%, well below the 50% reported by Lewis, Rogalski, and Seward (1999) and the 84% (2000-2008) and 65% (2009-2010) reported by Duca et al. (2012) for cross-industry U.S. samples. Dutordoir and Van de Gucht (2009) show that European convertibles tend to be structured more similarly to debt. They attribute this to differences in financial systems (banks versus capital markets), and they document a 27% average conversion probability. Despite our U.S. issuer tilt (60%), we find a considerably lower conversion probability. Our results are qualitatively insensitive to variations in the choice of government debt instrument or time periods used.

We find instead that low conversion probabilities are driven by high conversion premiums, and thus represent strong issuer growth assumptions. Therefore, based on our analysis of the attached call option, convertibles are priced to resemble debt rather than equity. The out-of-the-money calls make conversion into equity unlikely upon issuance in any case, and they are in line with the sweetened debt hypothesis (Brennan and Schwartz, 1988; Green, 1984; Mayers, 1998). Issuers face high costs of equity financing, triggered by high adverse selection costs, and by investors, who demand additional "lemon" premiums. Issuers may also try to signal firm quality by structuring a debt-like security, which could signal their unwillingness to share risk and return.

---

[32] Some convertible debt is callable by the issuer and/or putable by the investor, and these options are sometimes exercised prior to maturity. Hence, delta values may be affected by other options (Lewis, Rogalski, and Seward, 2003). For the sake of comparability, we follow standard procedures and neglect call features here.

This figure shows the conversion probability distribution of our renewable energy sample versus selected prior studies. We calculate probabilities based on Black and Scholes (1973) as in Lewis, Rogalski, and Seward (1999), using a country-specific five-year government bond as the risk-free rate benchmark. We also include a three-year maturity estimate for convertible preferred equity, as used by issuers in their financial statements.

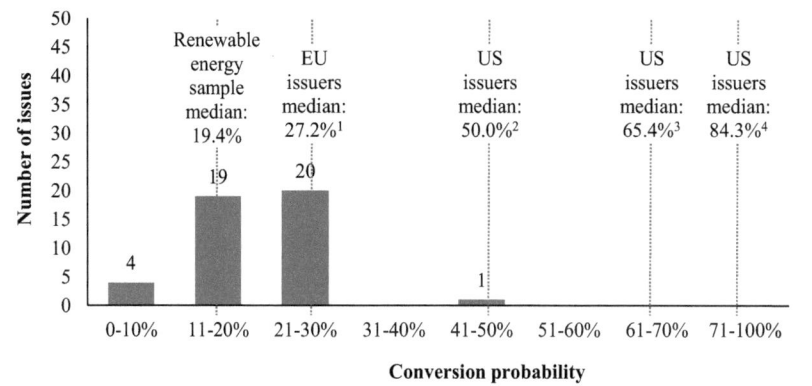

[1] Dutordoir and Van de Gucht (2009), sample period: 1994-2004, sample size: 179.
[2] Lewis, Rogalski, and Seward (1999), sample period: 1977-1984, sample size: 337.
[3] Duca et al. (2012), sample period: 2000-09/14/2008 (Lehman), sample size: 645, available from the authors upon request.
[4] Duca et al. (2012), sample period: 09/15/2008-2010, sample size: 64, available from the authors upon request.

Figure 4.1: CDO conversion probability distribution at issue

4.4 Short-term announcement effects

Given the debt-like structure of our sample and previous evidence that such security offerings are associated with less negative announcement returns, we expect any convertible issuer returns to be moderate and less negative than those of equity issuers. To test this hypothesis, we follow Brown and Warner (1985) in their widely used event study methodology, and add a three-factor model based on Fama and French (1992) and Faff (2003). We can thus minimize effects related to our limited sample size and heterogeneity in terms of issuer characteristics. Specifically, we calculate expected returns as:

$$E(R_i) = R_f + b_i[E(R_m) - R_f] + s_i E(SMB) + h_i (HML) \qquad (2)$$

where E(Ri) is the expected return on asset i, Rf is the return on the risk-free asset, E(Rm) is the expected return on the market portfolio, E(SMB) is the expected return on the mimicking portfolio for the "small minus large" size factor, and E(HML) is the expected return on the mimicking portfolio for the "high minus low" book-to-market factor.

Given the internationality of our sample, we use the following benchmarks to calculate the three factors: (i) the local market index, as suggested by Datastream, to account for regional market differences, (ii) the return differences between small and large companies, as well as (iii) companies with high and low market-to-book ratios, estimated using Frank Russell-style portfolios as proposed by Faff (2003). We use the following style indices to derive our factors: the Global Russell large-cap growth index (LCGI), Global Russell large-cap value index (LCVI), Global Russell small-cap growth index (SCGI), and Global Russell small-cap value index (SCVI). Specifically, we calculate the factors as:

$$SMB_t = \left(\frac{R_{SCVI_t} + R_{SCGI_t}}{2}\right) - \left(\frac{R_{LCVI_t} + R_{LCGI_t}}{2}\right) \quad (3)$$

$$HML_t = \left(\frac{R_{LCVI_t} + R_{SCVI_t}}{2}\right) - \left(\frac{R_{LCVI_t} + R_{SCGI_t}}{2}\right) \quad (4)$$

where SMBt and HMLt gauge the return differential between large/small companies and those with differing market-to-book ratios at time t. Finally, we also use returns from the one-month U.S. Treasury bill as a proxy for the risk-free interest rate. We regress the returns of the respective issuer against these factors over a 200-trading-day period ending ten days prior to the issue

This figure displays the time periods used to calculate abnormal returns. Expected returns are estimated over a 200 trading-day period, which ends ten days prior to the announcement of the financing event. The event window comprises 21 trading days.

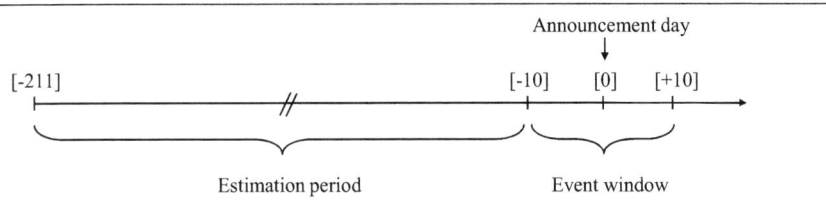

Figure 4.2: Event study estimation period and event window lengths

To test the statistical significance of our results, we use Boehmer, Musumeci, and Poulsen (1991) parametric test statistic, which adjusts for event-induced variance, as well as a non-parametric Wilcoxon test. Both have been shown to perform well for national and international samples (Campbell, Cowan, and Salotti, 2010). Table 4.2 gives our results.

Table 4.2: CDO and SEO short-term announcement effects

This table reports cumulative average abnormal returns (CAARs) for a set of 44 convertible debt and 285 seasoned equity offerings. We obtain issuer names from Bloomberg New Energy Finance (BNEF), transaction information from Thomson Reuters, and financial statement and market price information from Datastream. We calculate expected returns using a three-factor model based on Fama and French (1992) and Faff (2003). We estimate factors over a 200 trading-day period ending ten trading days prior to the issue. Factors are derived from global Russell portfolios and the respective market index of the issuer, as suggested by Datastream. We test for statistical significance of the results by using the Boehmer, Musumeci, and Poulsen (1991) parametric test statistic and the non-parametric Wilcoxon test (WCX). ***, **, and * denote significance levels at the 1%-, 5%-, and 10%-level, respectively.

	Convertible debt (n = 44)				SEOs (n = 285)			
Event window	CAAR Mean	Boehmer p-value	CAAR Median	WCX p-value	CAAR Mean	Boehmer p-value	CAAR Median	WCX p-value
[-10;+10]	-1.3%	(0.423)	-1.7%	(0.363)	-0.9%**	(0.111)	-2.0%**	(0.046)
[-5;+5]	-0.1%	(0.437)	-0.9%	(0.455)	-2.8%***	(0.001)	-2.8%***	(0.000)
[-1;+1]	-3.8%**	(0.038)	-2.1%**	(0.040)	-2.3%***	(0.000)	-2.2%***	(0.000)
[0;+1]	-4.6%**	(0.041)	-4.1%***	(0.006)	-1.9%***	(0.000)	-1.0%***	(0.000)

Our results show significantly negative announcement returns for both convertible debt and seasoned equity offerings. However, the mean abnormal return for convertible debt offerings is almost twice the size of that for seasoned equity offerings. This is contrary to the common finding that convertible debt is associated with less negative announcement returns than equity offerings (Dann and Mikkelson, 1984; Lewis, Rogalski, and Seward, 1999; Mikkelson and Partch, 1986). These studies relate the more moderate returns of CDOs with the signaling model of Myers and Majluf (1984), where equity offerings are perceived as stronger signals of firm overvaluations.

4.5 Issuer characteristics and failed signals

We next calculate key firm and market characteristics that have been shown to influence security design. Table 4.3 shows our calculations. Convertible issuers exhibit higher volatility, both stand-alone and when standardized by market volatility. This is in line with the notion that management will use higher volatility to reduce financing costs and mitigate the effects of asymmetric information.

Table 4.3: CDO and SEO market and firm characteristics at issue

This table provides descriptive statistics for our sample of 44 convertible debt issues and 285 seasoned equity offerings from 2001-2010. We obtain issuer names from Bloomberg New Energy Finance (BNEF), transaction information from Thomson Reuters, and financial statement and market price information from Datastream. *SIZE* is the issue volume converted into USD at the issue date. *RSIZE* is the issue volume divided by the market value of equity measured ten days prior to the issue. VOL_I is the annualized volatility over the interval from -211 to ten trading days prior to the issue. VOL_M is the volatility of the respective local market index over the same period. As per Datastream, we use market indices as benchmarks. *SPRUP* is the buy-and-hold return of the issuer over the same period. *NETRUP* is the issuer's buy-and-hold return net of the market's return. *SPRDOWN* and *NETRDOWN* are calculated similarly, starting with day 11 following the announcement. *EBETA* is the issuer's equity beta calculated against the local market index. *ABETA* is the unlevered equity beta under the assumption of a zero-debt beta. We use the latest fiscal year-end data for debt and the market value of equity measured ten trading days prior to the announcement. For the sake of comparability, we ignore any tax effects. We test differences in medians using a Wilcoxon test (WCX) and display corresponding p-values in parenthesis.

Variable	Convertible debt (n = 44)					SEOs (n = 285)					WCX
	Mean	Median	Stdev.	Max	Min	Mean	Median	Stdev.	Max	Min	p-value
SIZE	178	45	288	1,700	5	152	47	380	4,540	5	(0.599)
RSIZE	0.23	0.17	0.27	1.66	0.03	0.26	0.15	0.68	11.17	0.01	(0.710)
VOL_I	0.80	0.70	0.34	1.79	0.27	0.68	0.59	0.33	1.97	0.10	(0.031)
VOL_M	0.21	0.18	0.11	0.49	0.09	0.20	0.18	0.11	0.56	0.06	(0.820)
VOL_I/VOL_M	4.72	4.27	2.92	18.56	1.21	3.86	3.28	2.30	18.76	0.32	(0.072)
SPRUP	-0.04	-0.12	0.54	1.80	-0.84	0.52	0.21	1.19	8.94	-0.90	(0.000)
NETRUP	-0.04	-0.12	0.49	1.58	-0.79	0.46	0.18	1.12	8.72	-1.13	(0.000)
SPRDOWN	-0.16	-0.28	0.57	1.28	-0.91	-0.07	-0.16	0.55	2.51	-1.00	(0.371)
NETRDOWN	-0.13	-0.21	0.48	1.11	-0.99	-0.08	-0.14	0.47	1.82	-1.11	(0.538)
EBETA	1.18	1.18	0.59	2.20	-0.34	1.03	0.97	0.67	3.51	-0.26	(0.123)
ABETA	0.91	0.91	0.57	2.09	-0.34	0.85	0.82	0.60	2.69	-0.26	(0.537)

We find significant differences in stock price performance prior to issuance, which is negative for our convertible debt sample and strongly positive for our SEO sample. Previous cross-industry research has found positive run-ups of between 8.5% and 31.4% for both groups (Duca et al., 2012; Dutordoir and Van de Gucht, 2009; Lewis, Rogalski, and Seward, 2003). Note that issuing debt-like securities after a share price decline is at odds with conventional capital structure models, such as the pecking order model or the various tax and leverage cost trade-off models. This is because changes in the debt-to-equity ratio, triggered by a decline in the market value of equity, should lead to an equity-like issue in order to restore a firm's target capital structure.

We find no difference between the two groups with respect to size (absolute or relative) or risk (equity or asset beta). However, the post-issue performance of CDOs and SEOs is negative, on both an issuer and a net-of-market level. High volatility and weak relative stock price performance may lead issuers to provide market signals about firm quality by structuring more debt-like securities.

Table 4.4: CDO and SEO financial distress indicators at issue

This table provides financial distress-related statistics for our sample of 44 convertible debt issues and 285 seasoned equity offerings from 2001–2010. We obtain issuer names from Bloomberg New Energy Finance (BNEF), and transaction information from Thomson Reuters. Financial statement and market price information are based on Datastream. Except for the market value of equity, all ratios are based on the issuer's latest fiscal year-end financial statement data. *MTBV* is the company's market value of equity divided by the book value measured ten days prior to issue; *CA/TA* is the ratio of cash and marketable securities divided by total assets; *D/E* denotes issuers' debt-to-equity ratio; *LTD* represents issuers' long-term debt; and *OpCF*, *TCF*, and *EBITDA* are operating cash flow, total cash flow, and earnings before interest and taxes, depreciation, and amortization, respectively, each standardized by the interest expense (*IntEx*). We test differences in medians using a Wilcoxon test (WCX) and display corresponding p-values in parenthesis.

	Convertible debt				SEOs				WCX
	Median	Pos.	Neg.	n =	Median	Pos.	Neg.	n =	p-value
MTBV	2.06	81%	19%	42	2.04	97%	3%	279	(0.268)
CA/TA	0.22	100%	0%	40	0.18	100%	0%	282	(0.403)
D/E	0.51	100%	0%	44	0.38	100%	0%	285	(0.073)
LTD/TA	0.24	100%	0%	40	0.06	100%	0%	282	(0.002)
OpCF/IntEx	-2.61	35%	65%	37	-1.49	47%	53%	234	(0.439)
TCF/IntEx	18.76	84%	16%	37	18.91	85%	15%	234	(0.529)
EBITDA/IntEx	1.71	51%	49%	37	2.24	58%	42%	234	(0.703)
Altman z-score[1]	2.53	98%	3%	40	4.29	99%	1%	273	(0.022)

[1] Excluding the retained earnings indicator for availability reasons.

But, as we noted earlier, such signals are likely to fail. Stein (1992) and Kleidt and Schiereck (2009) argue that issuers who have been priced out of the equity markets may be forced to issue convertible debt to secure funding. This may occur, for example, if adverse selection costs render issuing equity prohibitively expensive. However, issuing convertibles may send a negative signal to the market that the issuer is unable to secure equity funding. We believe that deteriorating market value and high volatility prior to the issue are strong indicators of high equity-related costs of financing.

Signaling quality with a debt-like convertible may also be difficult if financing does not match the risk-return profile of the issuer or the industry. Debt-like instruments may be expensive in terms of fixed financing costs, and can repre-sent a drag on cash flow because the option (equity) component adds little to the total value of the security. We therefore calculate leverage and financial distress-related indicators for both groups in Table 4.4.

In line with the emerging industry nature of our sample, we find very low profitability and cash flow ratios. 49% of convertible issuers and 42% of seasoned equity issuers report negative earnings before interest and taxes. And more than half of both samples fail to meet their debt financing costs from their operating cash flow. Convertible debt issuers display significantly higher leverage ratios, both for market-related (debt equity) and book value-based indicators (long-term debt divided by total assets). Convertible issuers also have considerably lower Altman z-scores than equity issuers. We believe such financial distress indicators make it more difficult to establish quality signals via debt-like convertible securities, because they essentially add fixed capital servicing costs to cash-strapped issuers.

4.6 Cross-section analysis

We run cross-sectional regressions to confirm our results and to identify other drivers of abnormal returns. In particular, we seek to identify if temporal effects impact the magnitude of returns. If SEOs and CDOs are issued in grossly different time periods, comparisons of the two groups may be misleading. The financial crisis, for example, may constitute a time period of abnormal pricing and increased volatility. Any issues during this time period may be particularly badly received by the market. Our regression model incorporates a dummy variable for issues in the financial crisis and some of the volatility indicators from our previous analysis. We expect both to be negatively correlated with abnormal returns.

We further add a dummy variable which assumes the value of one if the company is from the solar industry and zero otherwise. We do so to capture a distinct industry effect. Solar companies were the most active issuer group in both CDO and SEO markets, and some have subsequently experienced financial distress (e.g. Q-Cells, Solon, Solyndra, Evergreen Solar). Any industry-specific impact may disturb our inferences with respect to security design.

We include other factors that have been shown to impact abnormal return as control variables. From our previous analysis, we expect leverage to negatively affect announcement returns. We also add variables for the stock price run-ups, relative issue size and relative valuation.

Table 4.5:   Cross-section regression results

This table gives the results of our cross-section regressions on abnormal returns for our sample of 44 convertible debt issues and 285 seasoned equity offerings in the period 2001–2010. The dependent variable equals the cumulative abnormal return (CAR) in the event window [-1;+1]. Regressions on other event windows show qualitatively similar results. We obtain issuer names from Bloomberg New Energy Finance (BNEF), and transaction information from Thomson Reuters. Financial statement and market price information are based on Datastream. *D/E* denotes issuers' debt-to-equity ratio *RSIZE* denotes issue proceeds divided by the market value of equity measured ten days prior to the issue. *NETRUP* is the issuer's buy-and-hold return net of the market's return over the -211 to ten trading days prior to announcement. $VOLA_I$ is the annualized volatility over the same period. $VOLA_M$ is the volatility of the respective local market index over the same period. *EBETA* is the ordinary least squares beta calculated over the same period. We use market indices as benchmarks, as suggested by Datastream. *CONV, CRISIS, SOLAR,* are dummy variables that indicate, respectively, whether the issue is a convertible debt issue, whether it was issued after Lehman Brothers' bankruptcy (09/15/2008), and whether the issuer is a member of the solar industry. We use White-adjusted standard errors to account for heteroscedasticity. ***, ** and * denotes statistical significance at the 1%-, 5%- and 10%-level respectively.

	Model 1 – CAR [-1+1]		Model 2 – CAR [-1;+1]	
	Coefficient	p-value	Coefficient	p-value
Constant	0.041***	(0.003)	0.044***	(0.002)
D/E	-0.001	(0.887)	-0.001	(0.867)
RSIZE	-0.002	(0.607)	-0.004	(0.437)
MTBV	-0.001*	(0.096)	-0.001	(0.118)
NETRUNUP	-0.009	(0.117)	-0.010*	(0.069)
$VOLA_I$	-0.021	(0.429)	-0.014	(0.595)
$VOLA_M$	-0.093	(0.164)	-0.105	(0.152)
EBETA	-0.021**	(0.027)	-0.016*	(0.093)
CONV			-0.014	(0.438)
CRISIS			-0.004	(0.751)
SOLAR			0.009	(0.454)
# of observations	321		321	
F-Value	4.390***	(0.000)	4.040***	(0.000)
R^2	0.070		0.090	

We estimate OLS-regressions and use robust standard errors (White, 1980) to account for heteroscedasticity in our sample. For robustness reasons, we estimate two models, either with or without dummy variables. We use the abnormal returns from the event window immediately surrounding the announcement (-1;+1), and our results remain qualitatively unchanged if we use other event periods.

The results shown in Table 4.5 are in line with our previous analysis. Except for the solar dummy variable, all indicators exhibit the expected sign. Leverage is negatively correlated with abnormal returns, as are relative valuation and risk indicators. We also conclude that our previous results are not significantly influenced by temporal or industry effects. Both the crisis and solar dummy remain statistically insignificant.

The convertible dummy displays the expected, negative sign, but also fails to be significant at the 10% level. Please note that this analysis seeks to control for other drivers of abnormal returns than security design. The fact that there is no significant difference between CDO and SEO abnormal returns supports our view that renewable energy convertible debt incurs different (relative) announcement returns than convertible debt in other (cross-industry) settings. However, the asymmetric sample sizes qualify this particular conclusion.

4.7 Risk dynamics and buy-side effects

4.7.1 Risk dynamics

We next analyze whether short-term announcement returns are driven by systematic risk changes. In their real option framework (for seasoned equity offerings), Carlson, Fisher, and Giammarino (2006) argue that systematic risk should decrease when a company replaces risky growth options with assets. Their model provides an explanation for the regularly reported risk-return pattern surrounding financing events, where risk and return tend to increase prior to an offering, and short- and long-term underperformance occur afterward (Ritter, 2003). According to their model, the prior risk increase is due to an increase in the growth option leverage as the option moves into the money and optimal investment progresses. The subsequent decline is triggered by a deleveraging of the growth option as the option is replaced by assets.

Zeidler, Mietzner, and Schiereck (2012) extend this framework to convertible bonds. Although they are hybrid in nature, convertible bonds represent additional financial leverage to firms at issuance. Additional leverage may cause increased agency conflicts between debt- and equityholders. Thus, better investment decisions can happen only "[...] if capital market investors value the convertibility option higher than the agency costs" (Zeidler, Mietzner, and Schiereck, 2012, p. 273). However, Zeidler, Mietzner, and Schiereck (2012) show that, after controlling for firm, risk, and market characteristics, equity and convertible issues behave similarly in that both display significant short- and long-term decreases in risk of similar magnitude.

Lewis, Rogalski, and Seward (2002) report systematic risk decreases after the issue of convertible bonds, a phenomenon they attribute to slow information processing by the market. Rai (2005) similarly shows a systematic risk decline after issuance, and argues that equity dilution and changes in leverage are the main drivers. Kleidt and Schiereck (2009) compare the risk dynamics of convertible bonds with equity issuers, and document a risk increase for convertibles only, which they attribute to some issuers being priced out of the equity markets.

For the sake of robustness, we estimate two models: 1) a three-factor model using the local market index as a benchmark (which corresponds to our event study model), and 2) to adjust for industry risk effects, a three-factor model using an industry index as a benchmark.

As Table 4.6 shows, beta changes are consistently negative for convertible debt issuers, and they fluctuate for seasoned equity offerings. All remain statistically insignificant. However, as Carlson, Fisher, and Giammarino (2010) note, investments and option exercise may not coincide with financing events, because the complexity of the projects may require investments over longer periods of time. Risk changes may thus also tend to materialize over longer periods of time. However, we are unable to test this hypothesis given the frequency of financing events in the industry and our limited sample size. On a short-term basis, we do not find that risk changes drive announcement returns.

Table 4.6: Risk changes surrounding CDOs and SEOs

This table reports pre- and post-issue betas for our sample of 44 convertible debt and 285 seasoned equity offerings. We calculate betas by using a three-factor model based on Fama and French (1992) and Faff (2003). As market benchmarks, we use both a local market index (panel A), and an industry index (i.e., Wilderhill New Energy Innovation Index, panel B). We run beta regressions over a 200-trading day period starting either 211 trading days before the issue or 11 days afterward. We test our samples for normal distributions using the Shapiro-Wilk test. P-values relate to standard t-tests (means) and Wilcoxon tests (medians).

		Convertible debt (n = 44)				SEOs (n = 285)			
		Beta before	Beta after	Beta change	p-value	Beta before	Beta after	Beta change	p-value
Panel A: Market index									
Equity beta	Mean	1.184	1.074	-0.109	(0.201)	1.032	1.053	0.022	(0.552)
	Median	1.183	1.045	-0.021	(0.113)	0.973	1.050	-0.011	(0.630)
Panel B: Industry index									
Equity beta	Mean	0.847	0.826	-0.020	(0.742)	0.655	0.664	0.009	(0.706)
	Median	0.770	0.789	-0.013	(0.430)	0.529	0.587	-0.001	(0.370)

4.7.2 Buy-side effects

Duca et al. (2012) argue that announcement returns of convertible bond issues are influenced by hedge fund activity. They hypothesize that "recent convertible bond issues may partly reflect temporary price pressure associated with arbitrage-induced short selling upon convertible bond issuance" (p. 1). Duca et al. (2012) create a proxy for hedge fund activity and relate it to the size of convertible bond announcement returns across three periods (1984-1999, 2000-09/14/2008, and 09/15/2008-2010 (post-Lehman)). Given the lack of data on renewable energy issuers prior to 2001, we are unable to fully replicate this analysis and compare announcement returns across the three periods. However, we believe our sample is affected less by buy-side effects, for several reasons.

First, the majority of our convertible debt sample is not publicly offered. As private placements, they allow for signals to be conveyed, but hedge funds, which often require a liquid market for assets and arbitrage transactions, are not able to invest in them. Second, in contrast to Duca et al. (2012), we find no significantly positive returns after offerings. If we observe price pressure from arbitrage, stock prices should quickly revert to previous levels. Our median average announcement return is 0.9% for period [+2;+5] and -1.2% for [+2;+10], which are both statistically insignificant. Third, we find no differences in announcement returns when we separate issues before and during the financial crisis, two periods with substantial differences in hedge fund activity.

4.8 Conclusion

This chapter argues in favor of a life-cycle component in the design and market impact of convertible debt securities. The flexibility in structuring convertible debt and its risk-mitigating features make it a valuable financing option for high-risk issuers. We analyze convertible debt offerings in the renewable energy industry, and show that it is structured to resemble debt rather than equity. Delta values average just 19%, well below previous cross-industry studies. At the same time, these convertibles incur significantly negative abnormal announcement returns, twice as large as those of seasoned equity offerings. Our results are contrary to the common finding of cross-industry studies that debt-like security offerings are associated with a less negative market impact.

We interpret our results to be failed signals by issuers of convertible debt in the renewable energy industry. Convertibles are issued after a stock price decline and in the presence of significant stock price volatility. These high adverse selection costs induce issuers to provide quality signals to the market. Debt-like securities can provide such signals by documenting the issuers' unwillingness to share future profits and risk.

However, we believe these signals may fail because the market perceives the security choice as representative of prohibitively high adverse selection costs. These may ultimately price issuers out of equity markets and force them to use convertible debt to secure funding. Our convertible debt sample displays similar risk (beta) and return (market-to-book) characteristics, but exhibits higher leverage and financial distress indicators, which would typically warrant additional equity rather than debt financing. We further show that industry profitability and cash flow are low, which also supports the use of equity instead of debt. More than 50% of issuers recorded negative operating cash flow figures.

Our results have important implications for both emerging industry companies and investors. For example, one problem is that the risk-mitigating features of convertible debt may not materialize if issuers fail to credibly signal firm quality to the markets. Furthermore, excessive growth assumptions and mismatches between project risk-return and financing costs may render it more difficult to create credible signals.

5 M&A-Success in the Renewable Energy Industry

The renewable energy industry has seen rapid growth over the last decade. Fuelled by growing concerns about climate change and dwindling fossil fuel resources, governments have put aggressive stimulus packages in place to grow new businesses and support established players in the industry. The use of solar, wind, hydro, geothermal and tidal energy sources as well as biomass is widely regarded as a key element of future energy supply. In 2008 more power capacity was added from renewable energy sources than from conventional means in both the European Union and the United States (REPN, 2009).

The growth of the industry has induced a wide range of companies, new ventures and established companies alike, to enter the market. Utility companies are buying into renewable energy to decrease their carbon footprint and diversify their energy sources; big industrials are increasing their exposure to benefit from the attractive growth rates associated with the sector; innovative technologies are being picked up by rivals to increase efficiency in business models that are yet largely dependent on government stimulus. These developments have led to a strong increase in mergers and acquisitions (M&A) activity. While the five years from 2000 to 2004 saw an aggregate of 216 deals, this number was recorded in 2008 alone. Wind and Solar account for the bulk of transactions, but activity in biogas and hydro has accelerated as business models mature.

In the light of this growing M&A-activity, this study examines the wealth effects of renewable energy industry transactions. Research on the general impact of M&A-transactions has produced rich evidence on the existence of wealth effects, particularly on the generous premiums enjoyed by target firm shareholders. Andrade, Mitchell, and Stafford (2001) in their study of 3,688 mergers find an average abnormal return to target shareholders of 16% over a three-day period around the transaction announcement. Reviewing 13 and 21 M&A-studies respectively, Campa and Hernando (2004) and (Bruner, 2002) support this result and conclude that the short-term announcement return ranges between 15% and 30%, depending on the observation period.

Empirical studies on acquirer returns provide more ambiguous results and have frequently been found to be zero or negative (Loughran and Vijh, 1997). This has often been associated with biased or opportunistic behavior of the acquiring firm's management. Roll's (1986) hubris theory suggests that managers are prone to overconfidence, leading to an overestimation of synergy potential and subsequent mispricing of transactions. Jensen (1986) argues that in the presence of free cash flows and asymmetric information, managers have incentives to squander resources instead of returning funds to shareholders. The corresponding rise in firm size is likely to be accompanied by an increase in the manager's prestige, salary and other nonmonetary private benefits.

Mitchell and Mulherin (1996) provide a less gloomy rational for acquirers to engage in M&A activity. They suggest that takeover activity is driven by – and often an appropriate response to – industry-wide shocks. They argue that due to the varying nature and impact of such shocks, e.g. technology innovations or regulation, the prospects of firms after a transaction may vary across industries and their respective environment. If this holds true then acquirer returns should differ across industries and might well add value to acquirer shareholders, at least in some sectors.

Since then, the number of industry-specific studies on transaction wealth effects has increased. Cumming and Johan (2006) and Beitel, Schiereck, and Wahrenburg (2004) provide evidence on the positive returns to acquirers for transactions in the financial services industry (insurance and banking, respectively). Gross and Lindstädt (2005) document positive acquirer returns for the automotive, media, telecom, financial services and pharma/chemical industries, albeit with a sample constructed from a relatively narrow and potentially upward-biased time period (January 1998–August 2001).

The following characteristics make renewable energy a particularly interesting industry to study wealth effects: energy is an enormous market at the beginning of a major structural change. This change is gathering speed as (i) scientists now generally agree that anthropogenic greenhouse gas emissions are very likely to be the main reason for global warming (IPCC, 2007) ; (ii) renewable energy business models mature and break even, making them attractive for corporate investors; (iii) major energy providers, industrials and car makers that have long betted on the gradual improvement of conventional technologies are now investing in renewable energy; and (iv) governments are aggressively driving the development of renewable energy to meet the growing energy demand, reduce supply side risks and create renewable energy hubs to support employment in their respective country. Assuming a long-term substitution of conventional energy resources and the presence of similar scale economies that have led to oligopolistic structures in many energy markets, the speed of growth could be a critical success factor in the emergence of new market leaders. If this holds true then external growth by means of acquisitions could send a positive signal on the prospects of the acquirer.

A second argument for the difference of the renewable energy sector is the "green premium" frequently reported by researchers. Klassen and McLaughlin (1996) find significant positive returns for strong environmental management and significant negative returns for weak environmental management as indicated by performance awards and crises, respectively. Derwall et al. (2005) show that a portfolio with high environmental scores outperformed a portfolio of firms with low scores by 6% per annum over the period 1997–2003.

Linking previous fund research to corporate finance, Chan (2009) analyses 372 "environmental-oriented" equity issuers consisting of ecologically and socially conscious companies as well as firms with strong corporate governance systems in the period of 1990–2007. Contrary to most cross-section studies on IPO- (Ritter, 1991) and SEO-performance (Loughran and Ritter, 1997), Chan (2009) finds statistically significant positive short- and long-term returns after controlling for size, book-to-market value, and momentum indicators. He suggests that "green" companies are perceived to be less prone to corporate social crises and environmental disasters and thus command a premium to non-green firms. If returns of ecologically friendly companies are indeed perceived as being more sustainable then the acquisition of such a company could convey information on the strategy of the acquirer and the goal of its management to produce sustainable returns. This effect could be particularly valuable to diversified acquirers not yet involved in renewable energy who could "spread" the effect over non-green assets (e.g. industrials).

Another interesting aspect of renewable energy is the presence of climate externalities. It relates to the fact that not all benefits of producing environmentally friendly energy can be captured by the shareholders and thus leads to a socially suboptimal pricing thereof. The European Commission through the Extern E Project has tried to quantify the true costs of electricity generation including the impact on human health, agriculture and ecosystems (Extern E, 2004). Applying a system that charges the heavy-emitters, or provides benefits to non-emitters, the costs of producing electricity from coal or oil would almost double. The results have contributed to the EU introducing a carbon emission trading system which aims at mitigating one aspect of the reported cost gap associated with conventional means of energy production. The superior returns of renewable energy could be interpreted as an increasing likelihood – with governments and many companies pushing for a structural change – that energy production will ultimately be priced closer to its "true" costs. Consolidation could produce more powerful industry players, which in turn could exert more influence on policy makers to change the pricing system. Consolidation would send a positive signal on the prospects of the industry, including the acquirer.

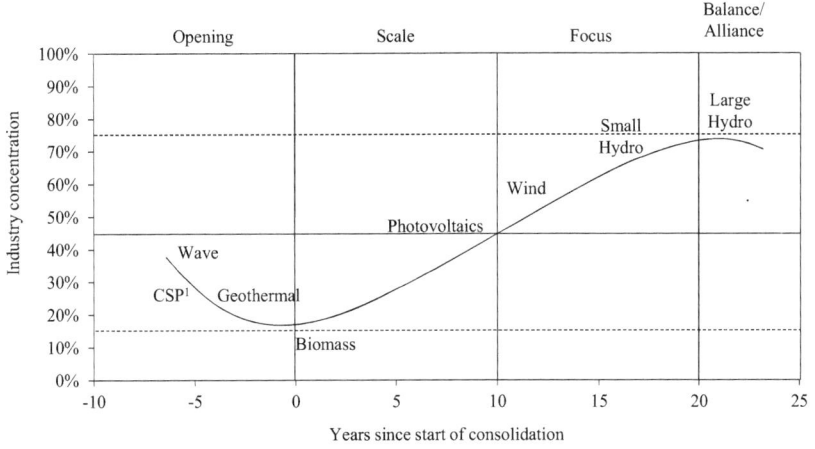

Source: A.T. Kearney (2008)
[1] Concentrated solar power (solar thermal)

Figure 5.1: Consolidation of renewable energy systems manufacturers by sub-sector

Finally, renewable energy is a heterogeneous industry offering insights into value effects of markets at different development stages and consolidation levels (see Figure 5.1 for an indicative categorization of consolidation levels by sub-sector). No single renewable energy source has evolved as the dominating carrier and it is now generally agreed that future renewable energy supply will most likely consist of a portfolio of technologies and sources. Wind and hydro were the earliest to enter the market and display the highest concentration levels. Solar and especially biomass are less concentrated and are expected to be subject to considerable M&A-activity in the coming years (A.T. Kearney, 2008). We expect acquirer gains to be higher (or losses lower) in less concentrated industries owing to the wider range of potential targets and non-transparent synergy gains, which tend to reduce the bargaining power of targets.

In our study, we analyze a total of 337 renewable energy transactions in the decade from 2000–2009 and focus on bidder returns. Our results show that (i) transactions in the renewable energy industry can increase acquiring shareholder wealth; (ii) the size of the acquirer and its market-to-book ratio negatively correlate with acquirer returns in the renewable energy industry, potentially indicating that shareholders assume discretionary spending and empire building by acquiring firm's management; (iii) firms from outside the industry can potentially gain a 'green premium' by buying into renewable energy, which is supported by the positive impact of such transaction on acquirer returns.

The remainder of the chapter is organized as follows: section 5.1 reviews related literature. In section 5.2, we discuss the sample selection procedure and the econometric approach. We present univariate and cross-sectional analyses of share price announcement effects to mergers and acquisitions in section 5.3 and 5.4, respectively. Section 5.5 concludes.

5.1 Literature review

For the related utility sector Becker-Blease et al. (2008), Berry (2000) and Bartunek et al. (1993) find large positive abnormal returns to target shareholders, significantly negative abnormal returns to acquirer shareholders and small but significant gains to the combined entity. Mentz and Schiereck (2008) document significantly positive abnormal returns to both the acquirer and the combined entity in cross-border transactions of the automotive industry. Acquisitions in high-tech industries, on the other hand, tend to have a negative wealth impact on acquirers, likely because the capital market perceives that bidders overpay for the growth potential (Ang and Kohers, 2001).

The majority of M&A studies focus on public targets, even though over 80% of all acquisitions involve a privately held company or a subsidiary (Fuller, Netter, and Stegemoller, 2002). There is evidence that the organizational form of the target plays an important role. Many of studies find that acquirers of private firms earn significantly positive abnormal returns, ranging from 1.3 to 2.1% (Capron and Shen, 2007; Chang, 1998; Draper and Paudyal, 2008; Fuller, Netter, and Stegemoller, 2002; Kohers, 2004; Moeller, Schlingemann, and Stulz, 2004). Since private companies are often small and not well-known, takeovers are less driven by adverse managerial incentives related to prestige and firm size (Draper and Paudyal, 2008). Private firms are often owned by only few individuals, and if acquirers pay with stock, the target owners form a large block in the merged company, which is likely to improve the monitoring of the acquirer's management. The willingness to accept equity without control also signals the quality of the buyer (Chang, 1998). Furthermore, the takeover market for private companies is typically illiquid and there is less bidder competition than for public assets. Target owners may thus have weaker bargaining power and accept lower premiums (Capron and Shen, 2007). This argument becomes particularly relevant when a company is too small to successfully operate on its own and seeks to be taken over by a more resourceful firm (Capron and Shen, 2007).

Hite et al. (1987) and Sichermann and Pettway (1992) focus on the organizational form of the target and analyze the divesture of subsidiaries, single divisions and other operating assets. In this case, they find significantly positive abnormal returns to both the seller and the acquirer. Fuller et al. (2002) and Moeller et al. (2004) document that buyers of subsidiaries earn even higher abnormal returns than acquirers of privately held firms. The authors stress that sellers increase their activity focus and that the proceeds of the sale can be invested in more profitable projects elsewhere (Hite, Owers, and Rogers, 1987). These gains might induce sellers to accept a lower price.

Besides the organizational form of the target, the size of the acquirer has an influence on the success of a merger. Moeller et al. (2004) show that small acquirers earn higher abnormal returns than large acquirers. This could be driven by higher free cash flows that encourage discretionary spending and empire building. It is more frequent in small companies that ownership and management is not separated, naturally motivating management to refrain from discretionary spending.

The size of the target relative to the acquirer also tends to affect abnormal returns. On one hand, the target needs to be sufficiently large to provide synergy potential and value impact to acquiring shareholders. On the other hand, large, complex targets may be difficult and costly to integrate. Here previous evidence provides ambiguous results. In a study of bank mergers, Beitel et al. (2004) report slightly higher abnormal returns to acquirers of relatively small targets. Focusing on private and subsidiary targets, both Fuller et al. (2002) and Draper and Paudyal (2006) find a positive relationship between relative size and abnormal returns to acquirers. In these cases, the higher value creation potential seems to compensate for the difficulties of integrating a more complex target.

The implications of geographical diversification on abnormal returns to acquirers have been regarded in several M&A studies. Engaging in cross-border deals provides access to new markets and growth opportunities. However, cultural and legal barriers might make it difficult to realize synergy gains (Campa and Hernando, 2004). The results suggest that acquirers benefit more by national, i.e. geographically focused mergers than in cross-border transactions (Becker-Blease, Goldberg, and Kaen, 2008; Campa and Hernando, 2004; DeLong, 2001).

5.2 Data and methodology

5.2.1 Data

Relevant transactions were identified using the Thomson Reuters database. We retrieve all M&A transactions in the period 01/01/2000 through 10/15/2009, which fall into one of 26 relevant SIC-codes. We then query company descriptions and deal synopses for a set of renewable energy keywords. The remaining transactions are cross-checked individually to correctly identify those involving a renewable energy target.[33] In a second step, we exclude all transactions that do not comply with the following sampling criteria: (i) we eliminate any transactions below USD 5 million to limit the effect of illiquid assets or marginal transactions; for similar reasons, we exclude issuers with infrequent trading; (ii) bidders are required to gain a majority stake in the target, either by an outright acquisition of more than 50% of voting rights or raising its stake beyond 50% in a transaction involving at least 25% thereof; (iii) finally, we require stock prices for either target or bidder to be available in the Datastream database for at least 273 trading days before the announcement of the transaction. This procedure reduces the sample to a total of 337 transactions. Table 5.1 and Table 5.2 provide descriptive sample characteristics.

The majority of transactions took place in Europe and North America. In more than half of the cases the acquirer was headquartered in the United States, Germany, Canada or Spain. Over two thirds of the deals are deemed "operational", i.e. include a target that owns or operates renewable energy assets. Transactions in the hydropower sector almost entirely belong to this group. The same is true

---

33 Keywords used: solar, sun, photo, wind, geotherm, hydro+power, hydro+elec, hydro+turbine, hydro+generation, bio+mass, bio+fuel, bio+diesel, ethanol+fuel, bioethanol, bio+energy, biogas, landfill+gas, biorefining, tidal, wave, renewable+energy, renewa-ble+power, renewable+fuel, alternative+energy and alternative+fuel.

for biomass, where over half of the acquisitions involve producers of biodiesel or ethanol. While the majority of wind power transactions are operational as well, there is also a significant number of "technology" deals defined as equipment or systems manufacturers.

Table 5.1: Sample M&A transactions by acquirer/target continent of origin

This table shows the geographical distribution of the 337 M&A transactions in the renewable energy industry in the period 2000–2009. Transactions and transaction data are from Thomson Reuters.

		Continent of acquirer						
		Europe	North America	South America	Asia	Australia	Africa	Total
Continent of target	Europe	125	18	1	4	6	-	154
	North America	20	77	1	5	5	-	108
	South America	5	3	7		1	-	16
	Asia	4	2	-	36		-	42
	Australia	1	3	-	1	11	-	16
	Africa	1		-	-	-	-	1
	Total	156	103	9	46	23	-	337

Table 5.2 also documents the importance of companies from outside the sector: almost 30% of acquirers are utilities that add renewable energy assets to their existing portfolio of conventional power plants. In about 10% of the cases, the acquisition is undertaken by pure financial investors. Only 33% of acquirers are renewable energy companies themselves, the large majority of which engaged in horizontal acquisition. Vertical deals are defined as those involving two companies on different stages of the value chain, for example a manufacturer of wind turbines and a wind park operator. Cases in which a renewable energy company purchases a target from another renewable energy sector, classified as a diversifying transaction, are very rare.

Only a small number of targets in the sample are publicly listed. The majority is either privately held or a subsidiary. The latter category encompasses single renewable power assets or divisions of the selling company.

Table 5.2: Sample M&A transaction characteristics

This table shows transaction characteristics of 337 M&A transactions in the renewable energy industry in the period 2000–2009. Transactions and transaction data are from Thomson Reuters. Deal type *Technology* includes manufacturers of systems and parts. *Operational* targets own and operate renewable energy assets. *Other* includes developers, installers of renewable power plants and systems, development rights owners and special advisory providers. We exclude deals from fuel cells, energy storage and smart grids subsectors.

Characteristics	Overall (n=337)	% of total	Solar (n=65)	Wind (n=153)	Hydro (n=59)	Geo (n=11)	Bio (n=41)	Other (n=8)
Transaction volume (USDm)								
Overall	68.1		29.4	61.3	101.5	16.6	59.9	105.5
Technology	33.1		31.0	37.6	-	-	5.0	-
Operational	74.0		11.6	73.8	117.8	53.31	70.0	105.5
Other	11.8		-	17.5	83.7	10.1	6.5	-
Target/deal type								
Technology	80	23.7%	46	27	3	1	3	-
Operational	225	66.8%	9	111	54	7	36	8
Other	32	9.5%	10	15	2	3	2	-
Acquirer industry								
Renewable Energy	112	33.2%	37	44	13	6	9	3
thereof horizontal	89		31	34	11	3	7	3
vertical	18		4	9	1	2	2	-
diversifying	5		2	1	1	1	-	-
Utility	96	28.5%	4	60	23	3	3	3
Financial Investor	29	8.6%	4	11	6	-	7	1
Other	100	29.7%	20	38	17	2	22	1
Public status of target								
Public	17	5.0%	3	4	3	-	3	4
Private	138	41.0%	37	62	14	6	19	-
Subsidiary	182	54.0%	25	86	42	5	20	4
Geographic focus								
National	168	49.9%	31	68	28	6	30	5
Intra-continent	88	26.1%	17	45	18	2	4	2
Cross-continent	81	24.0%	17	40	13	3	7	1

In about half of the transactions, acquirer and target are headquartered in the same country. The remaining half, in turn, splits up almost evenly into intra-continental and cross-continental transactions. In terms of geographic focus, the sub-sectors show approximately the same distribution as the complete sample, with the biomass sector representing an exception.

5.2.2 Methodology

We use a market model event study to assess the value effects to acquiring and target shareholders. This approach, based on (Brown and Warner, 1985), has been shown to work well in both national and international settings (Campbell, Cowan, and Salotti, 2010). For each company, we calculate expected returns with an OLS-regression over a period of 252 trading days (one full trading year) starting 273 trading days before and ending 20 trading days before the transaction announcement.

We use adjusted prices and industry/local market indices as suggested by Datastream. For robustness reasons, we also regress company returns against broader market indices, which resulted in qualitatively similar results (not reported). Together with market index information, the previously calculated beta can be used to derive expected returns for our target event window of 41 days surrounding the announcement of the transactions. We choose such a broad window to capture information leakage – common to M&A announcements – and allow for slow processing of information by the market. We then subtract actual company returns from expected returns to arrive at our measure of abnormal returns (CAR). We then aggregate CARs over the cross-section to calculate the average abnormal return (CAAR).

As frequently done and suggested by Harrington and Shrider (2007), we utilize the test statistic of Boehmer et al. (1991) to test the significance of cumulated abnormal returns. The test is to reflect a variance increase of abnormal returns in the event window. It includes a standardization of cumulated abnormal return following Mikkelson and Partch (1988). For the combined entity the adjusted standardization factor is calculated as proposed by Houston and Ryngaert (1994). In addition to these parametric tests, the non-parametric Wilcoxon test is employed to assess the statistical significance of median cumulated abnormal returns or percentages of positive cumulated abnormal returns, respectively. When comparing two groups of companies, we rely on a standard t-test as used, for example, by DeLong (2001).

5.3 Short-term announcement effects
5.3.1 Overall stock price performance

Figure 5.2 and Table 5.3 show the results of our analysis. For completeness reasons, we include target and combined entity returns despite their small sample sizes.

The average gain adds up to 1.6% in the [0;+1] event window and 2.2% in the [-5;+5] event window, both being statistically significant. Corresponding median CAARs are slightly lower, at 0.4% and 0.6%, respectively, but are also statistically significant. The returns to publicly traded renewable energy targets are consistent with findings in the M&A literature and shows that target shareholders earn a significant premium. On the announcement day alone, the excess return reaches 15.8%. Over the period of 41 days around the event day the average abnormal return sums up to a significant 27.2% (median 32.0 %).

This figure reports cumulated average abnormal returns (CAAR) to acquirers, targets and combined entities over a 41-day event window surrounding the announcement of a renewable energy transaction. Transactions and transaction data are from Thomson Reuters, share price data is from Datastream.

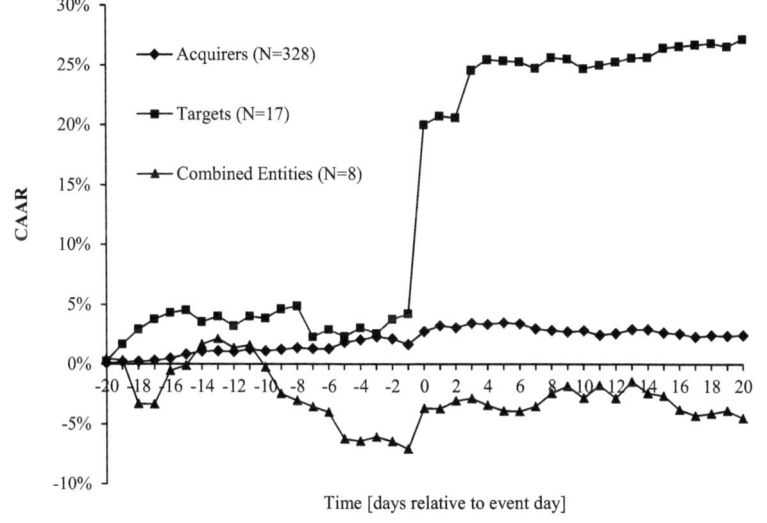

Figure 5.2: Short-term announcement effects of M&A transactions

Table 5.3: Short-term announcement effects of M&A transactions

This table reports mean average cumulated abnormal returns (CAAR) and median CAAR to acquirers (panel A), targets (panel B) and to the combined entities (panel C). CARs are estimated using OLS-regressions over a 252-day period and tested for statistical significance with a z-test as proposed by Boehmer, Musumeci, and Poulsen (1991). The statistical significance of the median CAAR is tested using the Wilcoxon test. % pos. gives the percentage of transactions resulting in a positive CAAR. ***, ** and * denotes statistical significance at the 1%-, 5%- and 10%-level respectively.

Event window	Mean CAAR	Boehmer p-value	Median CAAR	Wilcoxon p-value	% pos.
Panel A: Acquirers (n=328)					
[-20;-1]	1.64%	(0.540)	0.07%	(0.341)	50.6%
[-20;20]	2.42%	(0.216)	0.59%	(0.280)	52.1%
[-5;5]	2.19%*	(0.074)	0.57%*	(0.070)	55.2%
[-1;1]	1.10%**	(0.016)	0.38%**	(0.014)	55.5%
[-1;0]	0.62%	(0.108)	0.16%	(0.134)	53.0%
[0]	1.08%***	(0.005)	0.07%*	(0.057)	52.1%
[0;1]	1.56%***	(0.002)	0.38%***	(0.002)	57.3%
[1, 20]	-0.30%	(0.599)	-0.53%	(0.285)	47.6%
Panel B: Targets (n=17)					
[-20;-1]	4.2%*	(0.082)	3.54%	(0.210)	64.7%
[-20;20]	27.17%**	(0.026)	31.99%***	(0.003)	76.5%
[-5;5]	22.44%**	(0.042)	14.44%***	(0.002)	82.4%
[-1;1]	16.96%**	(0.022)	4.54%***	(0.009)	82.4%
[-1;0]	16.22%**	(0.047)	3.72%*	(0.068)	52.9%
[0]	15.78%**	(0.045)	2.51%*	(0.055)	64.7%
[0;1]	16.53%**	(0.018)	3.46%**	(0.013)	82.4%
[1;20]	7.17%	(0.226)	0.74%	(0.246)	52.9%
Panel C: Combined (n=8)					
[-20;-1]	-7.08%*	(0.070)	-4.41%	(0.148)	25.0%
[-20;20]	-4.50%	(0.370)	-5.93%	(0.313)	25.0%
[-5;5]	1.30%	(0.854)	-0.96%	(1.000)	50.0%
[-1;1]	2.73%	(0.131)	1.72%	(0.195)	75.0%
[-1;0]	2.77%	(0.152)	2.81%	(0.250)	62.5%
[0]	3.39%**	(0.058)	2.37%**	(0.039)	87.5%
[0;1]	3.35%*	(0.049)	2.12%*	(0.055)	75.0%
[1;20]	-0.82%	(0.487)	-1.46%	(0.742)	37.5%

More than 76% of the targets experience positive abnormal returns. This supports the notion that target shareholders can extract a significant portion of expected synergies. Combined entity results show positive results in the magnitude of 3.4% on the event day and 1.3% over the 11-day period [-5;+5]. However, the small sample size (eight) and the public status of both target and acquirer introduce considerably more heterogeneity. The negative returns before the event may be due to information leakage and investors' concerns about pricing of the transaction.

One of the key findings of our analysis is that acquiring shareholders do earn positive returns in renewable energy transactions. This is in line with our expectations, but runs counter to the results of many cross-industry studies, that predominantly report negative returns to acquirers (Andrade, Mitchell, and Stafford, 2001; Bruner, 2002).

We interpret the predominantly positive market reaction to renewable energy transactions as a sign for early consolidation benefits in a still fragmented industry. Acquirers may benefit from the vast growth options of renewable energy targets, which in turn may only be exploited with the help of acquirers. Our results also show that large parts of the expected synergy gains are allocated to target company shareholders.

5.3.2 Transaction returns by sub-sector

The heterogeneity of the renewable energy industry may cause wealth effects to accrue in some sub-sectors, where consolidation is perceived to be particularly beneficial. In order to detect possible differences, the sample is divided into solar, wind, hydro, bio and geothermal acquisitions. The short-term abnormal returns to acquirers are reported in Table 5.4.

On average, acquirers of solar companies nominally earn the highest returns with 11.5% over 41 days around the announcement. The high median CAAR of 7.6% in the same period is also found to be significant. The early share price reactions may indicate information leakage and transaction anticipation by the market. Despite the recent restructurings and corporate failures, solar remains a fragmented industry which is expected to further consolidate in the coming years. Overcapacities, in particular in Asia, and reduced government subsidies are likely to lead to further market share changes. At this stage, up to 2009, the market seems to award early consolidation moves and endorse the focus on the benefit of economies of scale. The abnormal returns in another fragmented industry - biomass - support this view. Acquirers in this sector experience a statistically significant two-day value gain of 4.33% at the announcement.

Table 5.4: Short-term announcement effects by renewable energy sub-sector

This table reports mean average cumulated abnormal returns (CAAR) and median CAAR to acquirers of renewable energy targets, differentiated by sub-sectors: solar (panel A), wind (panel B), hydro (panel C), biomass (panel D) and geothermal energy (panel E). CARs are estimated using OLS-regressions over a 252-day period and tested for statistical significance with a z-test as proposed by Boehmer, Musumeci, and Poulsen (1991). The statistical significance of the median CAAR is tested using the Wilcoxon test. % pos. gives the percentage of transactions resulting in a positive CAAR. ***, ** and * denotes statistical significance at the 1%, 5%- and 10%-level respectively.

Event window	Mean CAAR	Boehmer p-value	Median CAAR	Wilcoxon p-value	% pos.
Panel A: Solar (n=63)					
[-20;-1]	9.57%**	(0.043)	5.40%**	(0.017)	58.7%
[-20;20]	11.52%*	(0.062)	7.57%**	(0.022)	60.3%
[-1;1]	0.47%	(0.908)	0.59%	(0.642)	57.1%
[0]	2.54%	(0.174)	0.23%	(0.289)	55.6%
[0;1]	1.94%	(0.701)	-0.03%	(0.742)	49.2%
[1;20]	-0.59%	(0.933)	-0.34%	(0.805)	49.2%
Panel B: Wind (n=150)					
[-20;-1]	-0.27%	(0.895)	0.05%	(0.798)	50.7%
[-20;20]	-1.83%	(0.647)	0.12%	(0.467)	50.7%
[-1;1]	0.95%**	(0.027)	0.41%**	(0.047)	54.7%
[0]	0.11%**	(0.041)	0.03%	(0.518)	51.3%
[0;1]	1.09%***	(0.003)	0.34%**	(0.027)	59.3%
[1;20]	-1.67%	(0.973)	-0.68%*	(0.069)	46.7%
Panel C: Hydro (n=58)					
[-20;-1]	0.51%	(0.696)	-0.36%	(0.766)	48.3%
[-20;20]	0.60%	(0.781)	-0.67%	(0.978)	48.3%
[-1;1]	0.82%	(0.192)	0.51%**	(0.041)	63.8%
[0]	0.58%	(0.165)	0.30%*	(0.051)	56.9%
[0;1]	0.89%	(0.176)	0.58%***	(0.007)	63.8%
[1;20]	-0.49%	(0.615)	-1.32%	(0.450)	39.7%
Panel D: Biomass (n=41)					
[-20;-1]	-1.00%	(0.903)	-1.15%	(0.573)	41.5%
[-20;20]	7.64%	(0.230)	2.44%	(0.236)	53.7%
[-1;1]	3.87%	(0.249)	0.16%	(0.315)	53.7%
[0]	3.41%	(0.142)	0.15%	(0.160)	51.2%
[0;1]	4.33%*	(0.060)	0.22%	(0.188)	56.1%
[1;20]	5.23%*	(0.079)	1.31%	(0.274)	56.1%
Panel E: Geo (n=11)					
[-20;-1]	-2.69%	(0.291)	-1.57%	(0.465)	36.4%
[-20;20]	-2.07%	(0.841)	-3.18%	(0.577)	36.4%
[-1;1]	0.77%	(0.991)	-0.18%	(0.898)	36.4%
[0]	0.57%	(0.650)	-0.74%	(0.320)	18.2%
[0;1]	1.46%	(0.960)	1.79%	(0.147)	72.7%
[1;20]	0.06%	(0.284)	1.37%	(0.831)	54.5%

Wind and Hydro also show positive returns to acquirers, albeit at a lower level. Both experience an average return of approximately 1% (in the [0;+1] and [-1;+1] event window). Both markets are characterized by a higher level of consolidation and the predominant types of transactions are operational. Wind deals account for the largest number of transactions in the sample. They have led to the ten leading wind companies in 2008 accounting for a combined market share of 84% (BMU, 2009, p.14). Similarly, hydro, being the most established renewable energy technology in Europe, is dominated by three companies controlling approximately 55% of the market: Voith Hydro (Germany), Andritz Hydro (Austria) and Alstom Power Generation (France) (Gottwald, 2009, p. 5). One explanation for the lower acquirer returns compared to solar could be that, with greater experience, the value and synergy potentials of operating assets might be easier to asses. In this case, it is possible that owners of the target require a premium that distributes a large chunk of the benefits to them, leaving acquiring shareholders with only a small portion of the gain.

Drawing inferences on geothermal transactions is limited because of the small sample size. With this caveat in mind, the lack of any statistically significant abnormal returns suggests that, in the short-term, acquirers do not significantly benefit from M&A; at the same time, they do not suffer from the negative value effects so common in M&A-transactions.

5.3.3 Transaction returns by transaction characteristic

Besides industry affiliation and dynamics, many other factors have been shown to influence acquirer returns (Moeller, Schlingemann, and Stulz, 2005). To isolate their effect, we divide our sample according to transaction characteristics and compare abnormal returns.

Among the most consistent factors is the public status of the target. The acquisition of privately held and subsidiary targets has been found to be more beneficial to the shareholders of the acquirer than the takeover of publicly listed targets (Chang, 1998; Fuller, Netter, and Stegemoller, 2002; Hite, Owers, and Rogers, 1987; Mentz and Schiereck, 2008; Sicherman and Pettway, 1992). There is also evidence that the relative size of the target to the bidder matters (Asquith, Bruner, and Mullins Jr, 1983) in that larger targets relative to the size of the acquirer provide higher abnormal returns to acquiring shareholders (Asquith, Bruner, and Mullins Jr, 1983). Table 5.5 reports abnormal returns to these sub-groups of our sample

Table 5.5: Short-term announcement effects by transaction characteristics

This table reports mean average cumulated abnormal returns (CAAR) and value-weighted CAARs (VWCAAR) to acquirers of renewable energy targets, differentiated by transaction characteristics. The CAAR and the VWCAAR are tested for statistical significance with a z-test proposed by Boehmer et al. (1991) and a standard t-test, respectively. Group differences are tested with a standard t-test. ***, ** and * denotes statistical significance at the 1%-, 5%- and 10%-level respectively

Event window	Mean CAAR	Boehmer p-value	Mean VWCAAR	t-test p-value	% pos.
Panel A: Public status of target					
Public Target (n=8)					
[-20;20]	-1.51%	(0.852)	-5.50%	(0.161)	37.5%
[0;1]	4.15%*	(0.092)	3.94%	(0.318)	75.0%
Private Target (n=138)					
[-20;20]	5.19%	(0.227)	-0.23%	(0.898)	57.2%
[0;1]	2.49%	(0.115)	-0.49%	(0.323)	55.8%
Subsidiary target (n=182)					
[-20;20]	0.48%	(0.527)	1.65%	(0.208)	48.9%
[0;1]	0.74%**	(0.014)	0.43%	(0.233)	57.7%
Group difference (private target -subsidiary)		t-test p-value		t-test p-value	
[-20;20]	4.71%	(0.126)	-1.88%	(0.284)	-
[0;1]	1.75%	(0.104)	-0.92%	(0.121)	-
Panel B: Relative size of target					
High Relative Size[1]					
[-20; 20]	2.18%	(0.357)	-1.55%	(0.352)	52.0%
[0;1]	2.76%***	(0.005)	2.01%**	(0.012)	66.7%
Low Relative Size[2]					
[-20;20]	-2.83%**	(0.027)	-0.73%	(0.726)	41.3%
[0;1]	0.17%	(0.703)	0.17%	(0.767)	53.3%
Group difference (high-low)		t-test p-value		t-test p-value	
[-20,20]	5.01%*	(0.088)	-0.82%	(0773)	-
[0,1]	2.59%*	(0.056)	1.84%	(0.991)	-
Descriptive sample characteristics by sub-sector					
	Solar	Wind	Hydro	Bio	Geo
Acquirer Size (USD m)	21.2	23.6	16.6	9.5	13.7
Relative Target Size	100.92%	11.84%	24.83%	97.16%	27.42%

[1] Sub-sample of the 75 transactions with the highest target size relative to the acquirer (average size ratio: 72.5%; median size ratio: 16.7%).
[2] Sub-sample of the 75 transactions with the lowest target size relative to the acquirer (average size ratio: 0.9%; median size ratio: 0.7%).

All three target types show positive abnormal returns in the [0;+1] event window. Since there are no statistically significant differences in means of the sub-samples we conclude that, contrary to most cross-section evidence, the public status of the target does not influence acquirer returns in our renewable energy sample.

Regarding size there is weak evidence that relative transaction size does influence abnormal acquirer returns. Acquirers of relatively large targets on average earn 2.76% in the [0;+1] event window, highly significant at the 1% level. The means of both groups are significantly different for both the [0;+1] and [-20;+20] event window at the 10%-level. On a value-weighted basis, these differences turn insignificant. The potential reasons for such a difference include large synergy gains which outweigh integration costs or substantial advances in market position. Relative target size also seems to explain the large abnormal returns of the solar and biomass sub-samples. Table 5.5 reports the average acquirer size and relative target size by sub-sector of the renewable industry. On average, targets in solar (101% of acquirer, based on market capitalization 21 days prior to transaction and transaction value) and biomass (97%) by far outweigh targets in the wind (12%), hydro (25%) and geothermal (27%) sub-sectors.

5.4 Cross-section analysis

We now turn to cross-sectional regression and analyze the impact of multiple factors on abnormal returns. We estimate regressions using ordinary least squares and expand our set of factors to capture alternative drivers of acquirer returns. In addition to the previously reported variables (sub-sector, public status) we include factors for absolute acquirer size (logarithm of the market value 21 days prior to announcement), acquisition mode (dummy variable assuming 1 for indirect transactions or 0 for direct transactions), activity diversification (dummy variable assuming 1 if the acquirer is from outside the sector and 0 if otherwise), geographical diversification (dummy variables assuming 1 if the transaction is cross-boarder and 0 otherwise), market phase (dummy variable assuming 1 if transaction took place during the recent financial crisis, i.e. after Lehman filing for Chapter 11 and 0 otherwise) and priced growth and profitability (indicated by market-to-book ratio).

The results of the cross-sectional analysis are reported in Table 5.6. For robustness reasons, regressions are performed on the CARs of three different event windows. All three models are significant, the models on returns in the [-5;+5] and [-20;+20] event window at the 1%-level.

Table 5.6: Cross-section regression results

This table reports the results of our cross-sectional analysis on abnormal returns of acquirers in three event windows. *SOLAR, WIND, HYDRO* and *BIO* are industry dummy variables. *ACQUIRER SIZE* is the logarithm of the market value 21 days prior to announcement. *INDIRECT ACQUISITIONS* are transactions via a subsidiary of the acquirer. *DIVERSIFICATION, CROSS-BOARDER, PRIVATE TARGET, CRISIS* similarly are dummy variables which take on the value of one if the acquirer is from outside the industry, located in a different country or not publicly listed. *CRISIS* indicates whether the transaction took place after 08/15/2008. *MTBV* is the company's market-to-book value prior to the transaction. P-values relate to standard t-tests. ***, ** and * denotes statistical significance at the 1%-, 5%- and 10%-level respectively.

	Model 1 CAR [0;1]		Model 2 CAR [-5;+5]		Model 3 CAR [-20;+20]	
	Coefficient	p-value	Coefficient	p-value	Coefficient	p-value
Constant	0.024**	(0.013)	0.118**	(0.039)	0.163*	(0.059)
SOLAR	0.018	(0.301)	0.074	(0.120)	0.132*	(0.065)
WIND	0.014	(0.417)	0.030	(0.503)	-0.010	(0.880)
HYDRO	0.025	(0.555)	0.030	(0.538)	0.010	(0.888)
BIO	-0.020	(0.314)	0.054	(0.286)	0.054	(0.476)
ACQUIRER SIZE	-0.015***	(0.004)	-0.029***	(0.005)	-0.027*	(0.078)
INDIRECT ACQUISITION	0.031	(0.147)	-0.016	(0.444)	-0.003	(0.925)
DIVERSIFICATION	0.004***	(0.009)	0.028	(0.239)	0.017	(0.634)
CROSS-BOARDER	0.002	(0.709)	-0.020	(0.366)	-0.053	(0.105)
PRIVATE TARGET	-0.009	(0.860)	0.019	(0.341)	0.032	(0.283)
CRISIS	-0.001	(0.410)	-0.044*	(0.054)	-0.064**	(0.064)
MTBV	0.024	(0.603)	-0.009**	(0.034)	-0.018***	(0.005)
# of observations	321		321		321	
F-Value	2.731**	(0.022)	2.790***	(0.002)	3.360***	(0.000)
R^2	0.089		0.090		0.101	

The industry coefficients confirm the results from the previous section. Solar transactions during this time period were met with positive acquirer returns. It represents an endorsement of the economies of scale present in the industry and the benefits of consolidation, which continue beyond the sample period. We also note that the majority of positive returns accrue before the announcement date as indicated by differing significance levels across event windows. Information leakage may thus be a problem in solar transactions.

The most consistent impact on acquirer returns across all three models stems from the acquirer size variable. As Moeller et al. (2004) stress, managers of large companies may be more prone to overconfidence, and the availability of large free cash flows may encourage them to undertake negative net present value projects. Thus, motives other than shareholder value creation might drive acquisitions by large companies.

Significant diversification coefficients are consistent with the "green premium" argument described previously. By acquiring renewable energy assets, diversified companies can spread the sustainability signal over non-green assets. Transactions completed during the financial crisis yield significantly lower CAAR compared to pre-crisis deals. In this uncertain economic environment, investors may be especially risk averse and discount capacity expansions. Factors that endanger the realization of synergies may be overvalued.

Clearly contradicting the findings of Lang, Stulz, and Walkling (1989) and Servaes (1991) , a significant negative relation is found between short-term abnormal returns and the market-to-book value of the acquirer. Rau and Vermaelen (1998) document a similar relationship and argue that managers extrapolate the past performance of their company when considering mergers. Managers of firms that performed well, indicated by a high market-to-book value, may be overconfident when assessing the value creation potential of an acquisition.

5.5 Conclusion

In this study we investigate share price reactions to M&A-transactions in the rapidly growing renewable energy industry. We focus on acquirer effects and document positive returns on a sample of 337 transactions completed in the period 2000–2009. We read this as shareholder approval of a healthy consolidation in an emerging industry where larger players are more likely to extract efficiency gains and vast resources are required to successfully compete in the market. External growth is seen as an increased likelihood of being among the future winners of a still fragmented industry.

We document that acquirers from outside the industry earn positive abnormal returns when diversifying their activities towards renewable energy. We interpret this as a positive signal to shareholders about the sustainability of returns.

Diversified acquirers can spread this 'green premium' over non-green assets. We further show that the size of the acquirer and its book-to-market value correlate negatively with acquirer returns. This corresponds to the notion that large firms' acquisitions are more likely to be motivated by other reasons than shareholder wealth. These firms need to carefully select target firms and trade off external growth with profitability. There is also evidence that deals undertaken in the recent financial crisis have a negative effect on acquirer returns, which could be the result of increased risk aversion and discounted synergy gains.

Consistent with previous findings in industry and cross-section studies, targets are found to earn considerable premiums. On average, the abnormal return adds up to 27.2% during the 41 days around the announcement. Thus, the results suggest that value gains accrue to both parties involved in renewable energy acquisitions. No meaningful inferences can be drawn from the analysis of value-weighted combined entities due to small sample size. Most targets are either privately held or subsidiary companies, which is further indication of the emerging nature of the industry.

Parts of this chapter have been published as follows: Eisenbach, S., Ettenhuber, C., Schiereck D., & von Flotow, P. (2011). Consolidation in the Renewable Energy Industry and Bidders' M&A-Success. *Technology and Investment*, 2, 81-91.

6 Concluding Remarks and Outlook

This dissertation investigated the financing constraints and financing behavior of renewable energy companies. It identified industry-specific financing challenges and analyzed specific means of financing corporate growth. This final chapter summarizes the key results and points to potential future research in the field.

Chapter 2 provides a detailed review of industry-specific financing constraints. It serves as a basis to subsequent analyses and shows that unique market features create a challenging environment for renewable energy investors: first, the emerging life-cycle stage and speed of innovation creates informationally opaque markets. Second, the predominantly small- and medium-sized structure of the industry make funding more difficult, because fixed transaction costs weigh heavily on projects' profitability. Third, government intervention comes at the cost of increased regulatory risk that many investors are unwilling to take. Fourth, despite the introduction of quotas and feed-in tariffs, many markets remain at an external cost disadvantage to conventional energy means. Finally, the capital intensity and longevity of energy projects may impede the availability of capital to high-risk projects.

However, our review shows that these arguments, while frequently echoed in case studies, have varying degrees of empirical support. In fact, only the transaction costs-based underfunding has convincingly been shown to impede the (equity) financing of smaller companies. The lack of research in the field is due to both data availability as well as methodological difficulties. The latter derive from the fundamental problem of isolating good from bad investment projects as some may not receive adequate financing as a result of a bad business case, while others may suffer from market failure. Proxies that circumvent this problem, such as advanced financial statement analysis as conducted by Venturelli and Gualandri (2009), are likely to benefit from ever increasing data availability. The digitalization of the German Bundesanzeiger, for example, should offer rich opportunities for this angle to researching market failure.

The second study turns to publicly listed renewable energy companies and analyzes their financing behavior when issuing equity. It specifically investigates the impact of asymmetric information and growth opportunities on managers' decision to raise financing. It asks whether managers exploit the high level of asymmetric information to time the market by issuing equity when valuations are high. The presence of market timing would represent a strong indicator against the notion that renewable energy companies' development is impeded by financial constraints, at least for publicly listed companies.

We run logit regressions on 492 seasoned equity issues in the period 2000–2009. We find no systematically elevated market-to-book values or stock price run-ups for companies that issue equity. Rather, we find that the issue decision is driven by company age, which is in line with the notion that emerging companies are in need for capital to finance growth. This is also supported by our use-of-funds analysis, which shows that most issuers are in dire need for capital: more than half would have run out of cash and more than two thirds would have operated on below average cash balances without the SEO proceeds. Issuers spend heavily on capital expenditures and increase borrowing, both indicative of growth financing rather than market timing. Our results suggest that market timing is less pronounced in emerging industries, where growth opportunities outweigh the incentive to exploit asymmetric information to time the market.

Some of our findings may warrant further research. In particular, strongly negative returns following the offering are at odds with this growth model. We show that these cannot be explained by risk dynamics and believe that signaling and slow information processing by the market are unlikely sources of these returns. Furthermore, the reason behind the clustering of renewable energy offerings shortly after the IPO remains largely unexplained. Given the substantial transaction costs associated with the issue, returning quickly to the market appears to be an inefficient means to obtain financing.

Part three is dedicated to convertible debt securities. These cater to the capital intensity of the industry and its distinct risk characteristics. Convertible debt often serves as a hedge when the issuer's or industry's risk is difficult to estimate. We analyze the use, design and market impact of convertible debt in the renewable energy industry and compare it to seasoned equity offerings. It shows that convertible debt so far plays a minor role as a financing instrument. We identify 44 convertible debt and 285 seasoned equity issues in the period 2001–2010, the majority of which was issued in 2006 and 2007.

Convertible debt is structured to resemble debt rather than equity as aggressive growth assumptions render the conversion of the attached call option unlikely. Its announcement is not well received by the market: issuers experience a negative average abnormal short-term return of -4.6% over a two-day period. Debt-like convertibles are typically a signal of confidence, because issuers are unwilling to share future returns. In previous research, they have been shown to be associated with less negative announcement returns. Our analysis documents that convertible issuers underperform the general market and other renewable energy equity issuers prior and after the issue. They also exhibit significantly higher leverage and financial distress indicators. These results demonstrate the level of asymmetric information and its influence on the financing of these firms. Our results indicate that high-risk issuers try to provide a quality signal to the market by structuring a debt-like security. However, these signals tend to fail, mainly because markets interpret them as a sign of issuers being rationed out of equity markets.

This chapter in particular raises highly relevant and interesting follow-on research questions. Why do renewable energy companies make so little use of a financing instrument that can be flexibly structured to closely mirror the industry's characteristics? Are there differences in design and impact across technologies? And what impact do convertibles have on the risk of the issuers' debt instruments, in particular if issuers experience high levels of financial distress? These questions will be easier to investigate once sample sizes offer richer grounds for further analysis.

The final chapter targets wealth effects following M&A announcements. It centers on the question whether business combinations create value and which party to the transaction benefits most from it. Our analysis represents an indirect test of the size-related financing constraint described in Chapter 2. Business combinations typically provide the target with access to the – often larger – corporate resources of the bidder (e.g. marketing and sales, research and development, production), including financing. If renewable energy companies are considered financially constraint, the combination of a high-growth target with a resource-rich suitor may be beneficial for both parties.

We analyze 337 transactions involving a renewable energy target in the period 2000–2009. As expected, but in contrast to many cross-industry studies, we find significantly positive bidder returns upon announcement. In our view, this is indicative of the vast synergy potential available in the early consolidation phase of the industry. The rich premiums to target companies are in line with previous research. Our cross-sectional regression on abnormal returns reveals that acquirer size and high valuations correlate negatively with returns, highlighting the markets' doubt over discretionary spending by management. Finally, bidders from outside the industry tend to earn positive returns when buying into renewable energy, which we interpret as a sustainability signal to the market.

Further research could follow the value implications over the longer-term as opposed to our short-term perspective. The operating performance of these combinations may also offer interesting insights into the long-term success of renewable energy M&A.

There are other interesting renewable energy corporate finance questions beyond those mentioned above, and many of them so far remain unaddressed. This dissertation has largely focused on equity financing as a means to facilitate high-risk project development and innovation. It complements early industry research on initial public offerings and the long-run performance of renewable energy issuers by Chan (2009). However, the liabilities side of many renewable energy companies is dominated by debt. Financing constraints on debt markets may thus offer rich and highly relevant research opportunities. Furthermore, there is considerable scope for research on behavioral aspects and cognitive biases in renewable energy financing. There is a growing investor base which focuses on so-called sustainable investments and their involvement in renewable energy financing is growing.

On a more general level, the renewable energy industry offers a rare opportunity to study the financing of emerging industries. These have been argued to exert an above average influence on the structural renewal and growth of the economy. Our current understanding of the role of finance in emerging industry formation and development leaves room for further research.

Bibliography

A.T. Kearney. (2008). Renewable Energy Markets - Will the Boom Go Bust? Retrieved (14.06.2012) from www.atkearney.de/content/veroeffentlichungen.

Abhyankar, A., & Dunning, A. (1999). Wealth Effects of Convertible Bond and Convertible Preference Share Issues: An Empirical Analysis of the UK Market. Journal of Banking & Finance, 23(7), 1043-1065.

Abramowitz, M. (1956). Resource and Output Trends in the United States since 1870. American Economic Review, 46, 5-23.

Achleitner, A.-K., Braun, R., Bender, M., & Geidner, A. (2009). Community Development Venture Capital: Concept and Status Quo in Germany. International Journal of Entrepreneurship & Innovation Management, 9(4), 437-452.

Achleitner, A.-K., & Poech, A. (2004). Familienunternehmen und Private Equity: Die psychologischen Bruchstellen. Retrieved (16.8.2010) from www.familienunternehmen.de/media/public/pdf/studien.

Acs, Z., & Mueller, P. (2008). Employment Effects of Business Dynamics: Mice, Gazelles and Elephants. Small Business Economics, 30(1), 85-100.

Acs, Z. J., & Audretsch, D. B. (2005). Entrepreneurship, Innovation and Technological Change. Foundations and Trends in Entrepreneurship, 1(4), 1-49.

Acs, Z. J., & Gifford, S. (1996). Innovation of Entrepreneurial Firms. Small Business Economics, 8(3), 203-218.

Almeida, H., Campello, M., & Weisbach, M. S. (2004). The Cash Flow Sensitivity of Cash. The Journal of Finance, 59(4), 1777-1804.

Ammann, M., Fehr, M., & Seiz, R. (2006). New Evidence on the Announcement Effect of Convertible and Exchangeable Bonds. Journal of Multinational Financial Management, 16(1), 43-63.

Andrade, G., Mitchell, M., & Stafford, E. (2001). New Evidence and Perspectives on Mergers. Journal of Economic Perspectives, 15(2), 103-120.

Ang, J., & Kohers, N. (2001). The Take-Over Market for Privately Held Companies: the US Experience. Cambridge Journal of Economics, 25(6), 723.

Asquith, P., Bruner, R. F., & Mullins Jr, D. W. (1983). The Gains to Bidding Firms from Merger. Journal of Financial Economics, 11(1-4), 121-139.

Asquith, P., & Mullins Jr, D. W. (1986). Equity Issues and Offering Dilution. In J. Edwards & et al. (Eds.), Recent Developments in Corporate Finance. Cambridge, New York and Melbourne: Cambridge University Press.

Baker, M., & Wurgler, J. (2002). Market Timing and Capital Structure. Journal of Finance, 57(1), 1-32.

Bank of England. (2000). Finance for Small Firms: A Seventh Report. Retrieved (12.10.2011) from www.bankofengland.co.uk/publications/Documents/financeforsmallfirms/fin4sm09.pdf.

Bannock Consulting. (2001). Innovative Instruments for Raising Equity for SMEs in Europe. Retrieved (10.05.2011) from http://ec.europa.eu/enterprise/newsroom.

Baptista, R., Escária, V., & Madruga, P. (2008). Entrepreneurship, Regional Development and Job Creation: the Case of Portugal. Small Business Economics, 30(1), 49-58.

Bartunek, K., Jessell, K., & Madura, J. (1993). Are Acquisitions by Utility Firms Beneficial? Applied Economics, 25(11), 1401-1408.

Bator, F. M. (1958). The Anatomy of Market Failure. Quarterly Journal of Economics, 72(3), 351-379.

Bayless, M., & Chaplinsky, S. (1996). Is There a Window of Opportunity for Seasoned Equity Issuance? Journal of Finance, 51(1), 253-278.

Beatty, R. P., & Johnson, S. B. (1985). A Market-Based Method of Classifying Convertible Securities. Journal of Accounting, Auditing & Finance, 8(2), 112-124.

Becker-Blease, J., Goldberg, L., & Kaen, F. (2008). Mergers and Acquisitions as a Response to the Deregulation of the Electric Power Industry: Value Creation or Value Destruction? Journal of Regulatory Economics, 33(1), 21-53.

Beitel, P., Schiereck, D., & Wahrenburg, M. (2004). Explaining M&A Success in European Banks. European Financial Management, 10(1), 109-139.

Berger, A., & Udell, G. (1998). The Economics of Small Business Finance: The Roles of Private Equity and Debt Markets in the Financial Growth Cycle. Journal of Banking & Finance, 22(6-8), 613-673.

Berry, S. K. (2000). Excess Returns in Electric Utility Mergers During Transition to Competition. Journal of Regulatory Economics, 18(2), 175-188.

Black, B. S., & Gilson, R. J. (1998). Venture Capital and the Structure of Capital Markets: Banks versus Stock Markets. Journal of Financial Economics, 47(3), 243-277.

Black, F., & Scholes, M. (1973). The Pricing of Options and Corporate Liabilities. Journal of Political Economy, 81(3), 637.

Blanchflower, D. G., & Oswald, A. J. (1998). What Makes an Entrepreneur? Journal of Labor Economics, 16(1), 26.

Bloomberg New Energy Finance. (2010). Crossing the Valley of Death - Solutions to the Next Generation Clean Energy Project Financing Gap. Retrieved (09.08.2012) from www.bnef.com/WhitePapers/download/29.

Bloomberg New Energy Finance. (2012). Globale Renewable Energy Market Outlook with a Specific Focus on China. Retrieved (09.08.2012) from www.sinocleantech.com.

Bo, H., Huang, Z., & Wang, C. (2011). Understanding Seasoned Equity Offerings of Chinese Firms. Journal of Banking & Finance, 35(5), 1143-1157.

Boehmer, E., Musumeci, J., & Poulsen, A. B. (1991). Event-Study Methodology Under Conditions of Event-Induced Variance. Journal of Financial Economics, 30(2), 253-272.

Brennan, M., & Kraus, A. (1987). Efficient Financing Under Asymmetric Information. The Journal of Finance, 42(5), 1225-1243.

Brennan, M., & Schwartz, E. S. (1988). The Case for Convertibles. Journal of Applied Corporate Finance, 1(2), 55-64.

Brown, S. J., & Warner, J. B. (1985). Using Daily Stock Returns: The Case of Event Studies. Journal of Financial Economics, 14(1), 3-31.

Bruner, R. F. (2002). Does M&A Pay? A Survey of Evidence for the Decision-Maker. Journal of Applied Finance, 12(1), 48.

Bundesministerium für Umwelt, Naturschutz und Reaktorsicherheit. (2011). Erneuerbare Energien in Zahlen - Nationale und internationale Entwicklung. Retrieved (18.08.2012) from www.bmu.de/files/pdfs/allgemein/application/pdf/broschuere_ee_zahlen_bf.

Bundesministerium für Umwelt, Naturschutz und Reaktorsicherheit (Ed.). (2009). GreenTech made in Germany 2.0. München: Franz Wahlen.

Campa, J. M., & Hernando, I. (2004). Shareholder Value Creation in European M&As. European Financial Management, 10(1), 47-81.

Campbell, C. J., Cowan, A. R., & Salotti, V. (2010). Multi-Country Event-Study Methods. Journal of Banking & Finance, 34(12), 3078-3090.

Capron, L., & Shen, J. C. (2007). Acquisitions of Private vs. Public Firms: Private Information, Target Selection, and Acquirer Returns. Strategic Management Journal, 28(9), 891-911.

Carlson, M., Fisher, A., & Giammarino, R. (2006). Corporate Investment and Asset Price Dynamics: Implications for SEO Event Studies and Long-Run Performance. Journal of Finance, 61(3), 1009-1034.

Carlson, M., Fisher, A., & Giammarino, R. (2010). SEO Risk Dynamics. Review of Financial Studies, 23(11), 4026-4077.

Carpenter, R. E., & Petersen, B. C. (2002). Capital Market Imperfections, High-Tech Investment, and New Equity Financing. The Economic Journal, 112(477), F54-F72.

Chan, P. T. (2009). 'Green,' Are You Lovin It? An Examination of the Environmental-Friendly IPOs and SEOs. SSRN eLibrary, 1468902.

Chandler, A. D. (1954). Patterns of American Railroad Finance, 1830-50. The Business History Review, 28(3), 248-263.

Chang, S. (1998). Takeovers of Privately Held Targets, Methods of Payment, and Bidder Returns. Journal of Finance, 53(2), 773-784.

Cohen, S., Papadaki, A., & Siougle, G. (2007). SEOs in a 'Hot Market': Evidence of Timing. Applied Financial Economics, 17(14), 1179-1190.

Cornell, B., & Shapiro, A. C. (1988). Financing Corporate Growth. Journal of Applied Corporate Finance, 1(2), 6-22.

Cressy, R. (2002). Funding Gaps: A Symposium. The Economic Journal, 112(477), F1-16.

Cressy, R. (2012). Funding Gaps. In D. Cumming (Ed.), The Oxford Handbook of Entrepreneurial Finance. Oxford and New York: Oxford University Press.

Cumming, D., & Johan, S. (2006). Provincial Preferences in Private Equity. Financial Markets and Portfolio Management, 20(4), 369-398.

Dann, L. Y., & Mikkelson, W. H. (1984). Convertible Debt Issuance, Capital Structure Change and Financing-Related Information: Some New Evidence. Journal of Financial Economics, 13(2), 157-186.

Davidson, W. N., Glascock, J. L., & Schwarz, T. V. (1995). Signaling with Convertible Debt. Journal of Financial & Quantitative Analysis, 30(3), 425-440.

de Roon, F., & Veld, C. (1998). Announcement Effects of Convertible Bond Loans and Warrant-Bond Loans: An Empirical Analysis for the Dutch Market. Journal of Banking & Finance, 22(12), 1481-1506.

DeAngelo, H., DeAngelo, L., & Stulz, R. M. (2010). Seasoned Equity Offerings, Market Timing, and the Corporate Lifecycle. Journal of Financial Economics, 95(3), 275-295.

DeLong, G. L. (2001). Stockholder Gains from Focusing versus Diversifying Bank Mergers. Journal of Financial Economics, 59(2), 221-252.

Department of Business, Innovation and Skills. (2010). UK Innovation Investment Fund Launches £125m Environmental Investment Fund. Retrieved (14.05.2010) from www.bis.gov.uk/policies/innovation/businesssupport/ukiif.

Derwall, J., Guenster, N., Bauer, R., & Koedijk, K. (2005). The Eco-Efficiency Premium Puzzle. Financial Analysts Journal, 61(2), 51-63.

Devenow, A., & Welch, I. (1996). Rational Herding in Financial Economics. European Economic Review, 40(3-5), 603-615.

Draper, P., & Paudyal, K. (2008). Information Asymmetry and Bidders' Gains. Journal of Business Finance & Accounting, 35(3-4), 376-405.

Duca, E., Dutordoir, M., Veld, C., & Verwijmeren, P. (2012). Why are Convertible Bond Announcements Associated with Increasingly Negative Issuer Stock Returns? An Arbitrage-Based Explanation. Journal of Banking & Finance, forthcoming.

Dutordoir, M., & Van de Gucht, L. (2009). Why Do Western European Firms Issue Convertibles Instead of Straight Debt or Equity? European Financial Management, 15(3), 563-583.

Ebben, J., & Johnson, A. (2006). Bootstrapping in Small firms: An empirical Analysis of Change over Time. Journal of Business Venturing, 21(6), 851-865.

Eckbo, B. E. (1986). Valuation Effects of Corporate Debt Offerings. Journal of Financial Economics, 15(1–2), 119-151.

Eckbo, B. E., Masulis, R. W., & Norli, O. (2000). Seasoned Public Offerings: Resolution of the 'New Issues Puzzle'. Journal of Financial Economics, 56(2), 251-291.

Edler, D., Blazejczak, J., Wackerbauer, J., Rave, T., Legler, H., & Schasse, U. (2009). Beschäftigungswirkungen des Umweltschutzes in Deutschland: Methodische Grundlagen und Schätzung für das Jahr 2006. Retrieved (18.07.2011) from www.umweltdaten.de/-publikationen/fpdf-l/3846.pdf.

Eisenbach, S., Ettenhuber, C., Schiereck D., & von Flotow, P. (2011). Consolidation in the Renewable Energy Industry and Bidders' M&A-Success. Technology and Investment, 2, 81-91.

Ettenhuber, C., Schiereck, D., & von Flotow, P. (2011). Finanzierungsrestriktionen bei kleinen und mittleren Unternehmen der Umwelttechnologiebranche - Stand der Forschung und offene Fragen. Zeitschrift für Umweltpolitik und Umweltrecht, 34(1), 43-72.

European Commission. (2001). State Aid and Risk Capital. Official Journal of the European Communities, C 235, 3-11.

European Commission. (2006). Leitlinien der Gemeinschaft für staatliche Beihilfen zur Förderung von Risikokapitalinvestitionen in kleine und mittlere Unternehmen. Journal of the European Union, C 194, 2-21.

Evans, D. S., & Jovanovic, B. (1989). An Estimated Model of Entrepreneurial Choice under Liquidity Constraints. Journal of Political Economy, 97(4), 808.

Extern E. (2004). Externalities of Energy: Extension of Accounting Framework and Policy Applications. Retrieved (20.10.2010) from www.externe.info/externe_2006/expolwp3.pdf.

Faff, R. W. (2003). Creating Fama and French Factors with Style. Financial Review, 38(2), 311-322.

Fama, E. F. (1998). Market Efficiency, Long-Term Returns, and Behavioral Finance. Journal of Financial Economics, 49(3), 283-306.

Fama, E. F., & French, K. R. (1992). The Cross-Section of Expected Stock Returns. Journal of Finance, 47(2), 427-465.

Fama, E. F., & French, K. R. (2008). Dissecting Anomalies. Journal of Finance, 63(4), 1653-1678.

Fazzari, S. M., Hubbard, R. G., & Petersen, B. C. (1988). Financing Constraints and Corporate Investment. Brookings Papers on Economic Activity, 1, 141-195.

Fleming, P. D., & Probert, S. D. (1984). The Evolution of Wind Turbines: A Historical Review. Applied Energy, 18, 163-177.

Forbes. (2012). The World's Biggest Public Companies. Retrieved (12.07.2012) from www.forbes.com/lists/2006/18/Industry_14.html.

Fraser, S. (2004). Finance for Small and Medium-Sized Enterprises–A Report on the 2004 UK Survey of SME Finances. Retrieved (30.08.2010) from www.berr.gov.uk/files/file39407.pdf.

Fraunhofer ISE. (2012). Studie Stromgestehungskosten Erneuerbare Energien. Retrieved (10.08.2012) from www.ise.fraunhofer.de/de/veroeffentlichungen/veroeffentlichungen-pdf-dateien/studien-und-konzeptpapiere/studie-stromgestehungskosten-erneuerbare-energien.pdf.

Fritsch, M. (2008). Die Arbeitsplatzeffekte von Gründungen–Ein Überblick über den Stand der Forschung. Zeitschrift fur Arbeitsmarktforschung, 41(1), 55-69.

Fritsch, M., & Müller, P. (2008). The Effect of New Business Formation on Regional Development Over Time: The Case of Germany. Small Business Economics, 30(1), 15-29.

Fritsch, M., & Schilder, D. (2007). Is There a Regional Equity Gap for Innovative Start-Ups? The Case of Germany. In M. Dowling & J. Schmude (Eds.), Empirical Entrepreneurship in Europe: New Perspectives. Cheltenham and Northampton: Elgar.

Fuller, K., Netter, J., & Stegemoller, M. (2002). What Do Returns to Acquiring Firms Tell Us? Evidence from Firms That Make Many Acquisitions. Journal of Finance, 57(4), 1763-1793.

Gerhard, M., Rüschen, T., & Sandhövel, A. (Eds.). (2011). Finanzierung Erneuerbarer Energien. Frankfurt: Frankfurt School Verlag.

Goldfarb, B., Kirsch, D., & Shen, A. (2012). Finance of New Industries. In D. Cumming (Ed.), The Oxford Handbook of Entrepreneurial Finance. Oxford and New York: Oxford University Press.

Gompers, P., & Lerner, J. (2001). The Venture Capital Revolution. The Journal of Economic Perspectives, 15(2), 145-168.

Gompers, P., & Lerner, J. (2003). Money Chasing Deals? The Impact of Fund Inflows on Private Equity Valuations. In M. Wright, H. J. Sapienza & L. W. Busenitz (Eds.), Venture Capital. Elgar Reference Collection. Cheltenham Northampton: Elgar.

Gompers, P., & Lerner, J. (2004). The Venture Capital Cycle. Cambridge and London: MIT Press.

Green, R. (1984). Investment Incentives, Debt, and Warrants. Journal of Financial Economics, 13(1), 115-136.

Greenwald, B., Stiglitz, J. E., & Weiss, A. (1984). Informational Imperfections in the Capital Market and Macroeconomic Fluctuations. American Economic Review, 74(2), 194.

Griliches, Z. (1992). The Search for R&D Spillovers. Scandinavian Journal of Economics, 94, S29-S47.

Gross, S. K. H., & Lindstädt, H. (2005). Horizontal and Vertical Takeover and Sell-off Announcements: Abnormal Returns Differ by Industry. Corporate Ownership & Control, 3(2), 23-30.

H.M. Treasury Small Business Service. (2003). Bridging The Finance Gap-A Consultation on Improving Access to Growth Capital for Small Businesses. Retrieved (25.11.2011) from www.bis.gov.uk/assets/biscore/enterprise/docs/s/12-539-sme-access-external-finance.

Harding, R. (2000). Venture Capital and Regional Development: Towards a Venture Capital 'System'. Venture Capital, 2(4), 287-311.

Harding, R., & Cowling, M. (2006). Points of View–Assessing the Scale of the Equity Gap. Journal of Small Business and Enterprise Development, 13(1), 115-132.

Harjoto, M., & Garen, J. (2003). Why Do IPO Firms Conduct Primary Seasoned Equity Offerings? Financial Review, 38(1), 103-125.

Harrington, S. E., & Shrider, D. G. (2007). All Events Induce Variance: Analyzing Abnormal Returns When Effects Vary across Firms. Journal of Financial & Quantitative Analysis, 42(1), 229-256.

Hemer, J., Berteit, H., Walter, G., & Göthner, M. (2006). Erfolgsfaktoren für Unternehmensausgründungen aus der Wissenschaft. Stuttgart: Fraunhofer IRB.

Hite, G. L., Owers, J. E., & Rogers, R. C. (1987). The Market for Interfirm Asset Sales: Partial Sell-Offs and Total Liquidations. Journal of Financial Economics, 18(2), 229-252.

Hofman, D. M., & Huisman, R. (2012). Did the Financial Crisis Lead to Changes in Private Equity Investor Preferences Regarding Renewable Energy and Climate Policies? Energy Policy, 47(0), 111-116.

Houston, J. F., & Ryngaert, M. D. (1994). The Overall Gains from Large Bank Mergers. Journal of Banking & Finance, 18(6), 1155-1176.

Howe, J. S., & Zhang, S. (2010). SEO Cycles. Financial Review, 45(3), 729-741.

Hubbard, R. G. (1998). Capital Market Imperfections and Investment. Journal of Economic Literature, 36(1), 193-225.

Ibbotson, R. G., & Jaffe, J. F. (1975). "Hot Issue" Markets. The Journal of Finance, 30(4), 1027-1042.

Intergovernmental Panel on Climate Change. (2007). Climate Change 2007: The Physical Science Basis-Summary for Policymakers. Contribution of Working Group I to the Fourth Assessment Report of the Intergovernmental Panel on Climate Change. Retrieved (02/25/2010) from www.ipcc.ch/pdf/assessment-report/ar4/wg1/ar4-wg1-spm.pdf.

International Energy Agency. (2012). Key World Energy Statistics. Retrieved (02.08.2012) from www.iea.org/textbase/nppdf/free/2011/key_world_energy_stats.pdf.

Jegadeesh, N. (2000). Long-Term Performance of Seasoned Equity Offerings: Benchmark Errors and Biases in Expectations. Financial Management, 29(3), 5.

Jensen, M. C. (1986). Agency Costs of Free Cash Flow, Corporate Finance, and Takeovers. American Economic Review, 76(2), 323.

Jung, K., Kim, Y.-C., & Stulz, R. M. (1996). Timing, Investment Opportunities, Managerial Discretion, and the Security Issue Decision. Journal of Financial Economics, 42(2), 159-185.

Jung, M., & Sullivan, M. J. (2009). The Signaling Effects Associated with Convertible Debt Design. Journal of Business Research, 62(12), 1358-1363.

Kaldellis, J. K., & Zafirakis, D. (2011). The Wind Energy (R)Evolution: A Short Review of a Long History. Renewable Energy, 36(7), 1887-1901.

Kang, J.-K., Kim, Y.-C., Park, K.-J., & Stulz, R. M. (1995). An Analysis of the Wealth Effects of Japanese Offshore Dollar-Denominated Convertible and Warrant Bond Issues. Journal of Financial & Quantitative Analysis, 30(2), 257-270.

Kang, J.-K., & Stulz, R. M. (1996). How Different is Japanese Corporate Finance? An Investigation of the Information Content of New Security Issues. Review of Financial Studies, 9(1), 109.

Kaplan, S. N., & Zingales, L. (1997). Do Investment-Cash Flow Sensitivies Provide Useful Measures of Financing Constraints? Quarterly Journal of Economics, 112(1), 169-215.

Kenny, M. (2011). Venture Capital Investment in Greentech Industries: A Provocative Essay. In R. Wüstenhagen & R. Wübker (Eds.), Handbook Of Research On Energy Entrepreneurship. Cheltenham, UK and Northampton, MA: Edward Elgar Publishing Ltd.

Kim, W., & Weisbach, M. S. (2008). Motivations for Public Equity Offers: An International Perspective. Journal of Financial Economics, 87(2), 281-307.

Kim, Y. O. (1990). Informative Conversion Ratios: A Signalling Approach. Journal of Financial & Quantitative Analysis, 25(2), 229-243.

Klassen, R. D., & McLaughlin, C. P. (1996). The Impact of Environmental Management on Firm Performance. Management Science, 42(8), 1199-1214.

Kleidt, B., & Schiereck, D. (2009). Systematic Risk Changes Around Convertible Debt Offerings: A Note on Recent Evidence. Global Finance Journal, 20(1), 98-105.

Knight, E. R. (2010). The Economic Geography of Clean Tech Venture Capital. SSRN eLibrary, 1588806.

Kohers, N. (2004). Acquisitions of Private Targets: The Unique Shareholder Wealth Implications. Applied Financial Economics, 14(16), 1151-1165.

Korajczyk, R. A., Lucas, D. J., & McDonald, R. L. (1991). The Effect of Information Releases on the Pricing and Timing of Equity Issues. Review of Financial Studies, 4(4), 685-708.

Kuhlman, B. R., & Radcliffe, R. C. (1992). Factors Affecting The Equity Price Impacts Of Convertible Bonds. Journal of Applied Business Research, 8(4), 79-86.

Laffont, J. J. (2008). Externalities. In S. N. Durlauf & L. E. Blume (Eds.), The New Palgrave Dictionary of Economics. Basingstoke: Palgrave Macmillan.

Lang, L. H. P., Stulz, R. M., & Walkling, R. A. (1989). Managerial Performance, Tobin's Q, and the Gains from Successful Tender Offers. Journal of Financial Economics, 24(1), 137-154.

Lawton, T. C. (2002). Missing the Target: Assessing the Role of Government in Bridging the European Equity Gap and Enhancing Economic Growth. Venture Capital, 4(1), 7-23.

Lerner, J. (2002a). Boom and Bust in the Venture Capital Industry and the Impact on Innovation. Federal Reserve Bank of Atlanta Economic Review, 87(4), 25-39.

Lerner, J. (2002b). When Bureaucrats Meet Entrepreneurs: The Design of Effective 'Public Venture Capital' Programmes. The Economic Journal, 112(477), F73-F84.

Lerner, J. (2009). Boulevard of Broken Dreams: Why Public Efforts to Boost Entrepreneurship and Venture Capital Have Failed–and What to Do About It. Princeton and Woodstock: Princeton University Press.

Levin, R. C., Klevorick, A. K., Nelson, R. R., & Winter, S. G. (1987). Appropriating the Returns from Industrial Research and Development. Brookings Papers on Economic Activity, 1987(3), 783-831.

Levine, R. (2005). Finance and Growth: Theory and Evidence. In P. Aghion & S. N. Durlauf (Eds.), Handbook of Economic Growth. Amsterdam, London, Oxford and San Diego: North-Holland.

Lewis, C. M., Rogalski, R. J., & Seward, J. K. (1999). Is Convertible Debt a Substitute for Straight Debt or for Common Equity? Financial Management, 28(3), 5-27.

Lewis, C. M., Rogalski, R. J., & Seward, J. K. (2002). Risk Changes Around Convertible Debt Offerings. Journal of Corporate Finance, 8(1), 67-80.

Lewis, C. M., Rogalski, R. J., & Seward, J. K. (2003). Industry Conditions, Growth Opportunities and Market Reactions to Convertible Debt Financing Decisions. Journal of Banking & Finance, 27(1), 153-181.

Loughran, T., & Ritter, J. R. (1995). The New Issues Puzzle. Journal of Finance, 50(1), 23-51.

Loughran, T., & Ritter, J. R. (1997). The Operating Performance of Firms Conducting Seasoned Equity Offerings. Journal of Finance, 52(5), 1823-1850.

Loughran, T., & Vijh, A. M. (1997). Do Long-Term Shareholders Benefit From Corporate Acquisitions? Journal of Finance, 52(5), 1765-1790.

Lund, P. D. (2009). Effects of Energy Policies on Industry Expansion in Renewable Energy. Renewable Energy, 34(1), 53-64.

Marshall, A. (1920). The Principles of Economics; An Introductory Volume. London: Macmillan.

Martin, R., Berndt, C., Klagge, B., & Sunley, P. (2005). Spatial Proximity Effects and Regional Equity Gaps in the Venture Capital Market: Evidence from Germany and the United Kingdom. Environment and Planning A, 37(7), 1207-1231.

Mason, C. M., & Harrison, R. T. (1999). Public Policy and the Development of the Informal Venture Capital Market: UK Experience and Lessons for Europe. In K. Cowling (Ed.), Industrial Policy in Europe: Theoretical perspectives and practical proposals. London and New York: Routledge.

Mason, C. M., & Harrison, R. T. (2003). Closing the Regional Equity Gap? A Critique of the Department of Trade and Industry's Regional Venture Capital Funds Initiative. Regional Studies, 37(8), 855-868.

Masulis, R. W., & Korwar, A. N. (1986). Seasoned Equity Offerings: An Empirical Investigation. Journal of Financial Economics, 15(1-2), 91-118.

Mayers, D. (1998). Why Firms Issue Convertible Bonds: The Matching of Financial and Real Investment Options. Journal of Financial Economics, 47(1), 83-102.

Mentz, M., & Schiereck, D. (2008). Cross-Border Mergers and the Cross-Border Effect: The Case of the Automotive Supply Industry. Review of Managerial Science, 2(3), 199-218.

Merton, R. C. (1974). On the Pricing of Corporate Debt: The Risk Structure of Interest Rates. The Journal of Finance, 29(2), 449-470.

Meza, D. D., & Webb, D. C. (1987). Too Much Investment: A Problem of Asymmetric Information. The Quarterly Journal of Economics, 102(2), 281-292.

Mikkelson, W. H., & Partch, M. M. (1986). Valuation Effects of Security Offerings and the Issuance Process. Journal of Financial Economics, 15(1-2), 31-60.

Mikkelson, W. H., & Partch, M. M. (1988). Withdrawn Security Offerings. Journal of Financial & Quantitative Analysis, 23(2), 119-133.

Mitchell, M. L., & Mulherin, J. H. (1996). The Impact of Industry Shocks on Takeover and Restructuring Activity. Journal of Financial Economics, 41(2), 193-229.

Moeller, S. B., Schlingemann, F. P., & Stulz, R. M. (2005). Wealth Destruction on a Massive Scale? A Study of Acquiring-Firm Returns in the Recent Merger Wave. Journal of Finance, 60(2), 757-782.

Moeller, S. B., Schlingemann, F. P., & Stulz, R. M. (2004). Firm Size and the Gains from Acquisitions. Journal of Financial Economics, 73(2), 201-228.

Müller, P., Stel, A. v., & Storey, D. J. (2008). The Effects of New Firm Formation on Regional Development Over Time: The Case of Great Britain. Small Business Economics, 30(1), 59-71.

Myers, S. C., & Majluf, N. S. (1984). Corporate Financing and Investment Decisions When Firms Have Information That Investors Do Not Have. Journal of Financial Economics, 13(2), 187-221.

North, D., Smallbone, D., & Vickers, I. (2001). Public Sector Support for Innovating SMEs. Small Business Economics, 16(4), 303.

Oakey, R. (2003). Technical Entreprenenurship in High Technology Small Firms: Some Observations on the Implications for Management. Technovation, 23(8), 679-688.

Oakey, R. (2007). A Commentary on Gaps in Funding for Moderate 'Non-Stellar' Growth Small Businesses in the United Kingdom. Venture Capital, 9(3), 223-235.

OECD. (2006). The SME Financing Gap Theory and Evidence (Vol. 1). Paris: OECD.

Olmos, L., Ruester, S., & Liong, S.-J. (2012). On the Selection of Financing Instruments to Push the Development of New Technologies: Application to Clean Energy Technologies. Energy Policy, 43, 252-266.

Opler, T., Pinkowitz, L., Stulz, R. M., & Williamson, R. (1999). The Determinants and Implications of Corporate Cash Holdings. Journal of Financial Economics, 52(1), 3-46.

Paul, S., Rödel, S., Rödl, H., & Stein, S. (2005). Aufbruch aus dem Jammertal. Eigenkapitallücke und Eigenkapitalstrategien im Mittelstand. Creditreform Wissen, 9, 4-17.

Petersen, M. A. (2009). Estimating Standard Errors in Finance Panel Data Sets: Comparing Approaches. Review of Financial Studies, 22(1), 435-480.

Pleschak, F., & Werner, H. (1998). Technologieorientierte Unternehmensgründungen in den neuen Bundesländern. Heidelberg: Physica.

Popp, D. (2006). R&D Subsidies and Climate Policy: Is There a "Free Lunch"? Climatic Change, 77(3-4), 311-341.

Powell, W. W., Koput, K. W., Bowie, J. I., & Smith-Doerr, L. (2002). The Spatial Clustering of Science and Capital: Accounting for Biotech Firm–Venture Capital Relationships. Regional Studies, 36(3), 291-305.

Praag, C. M. v., & Versloot, P. H. (2008). The Economic Benefits and Costs of Entrepreneurship: A Review of the Research. Foundations and Trends in Entrepreneurship, 4(2), 65-154.

Rai, A. (2005). Changes in Risk Characteristics of Firms Issuing Hybrid Securities: Case of Convertible Bonds. Accounting & Finance, 45(4), 635-651.

Rajan, R. G., & Zingales, L. (1998). Financial Dependence and Growth. American Economic Review, 88(3), 559-586.

Rau, P. R., & Vermaelen, T. (1998). Glamour, Value and the Post-Acquisition Performance of Acquiring Firms. Journal of Financial Economics, 49(2), 223-253.

Renewable Energy Policy Network for the 21st Century. (2009). Renewables Global Status Report: 2009 Update. Retrieved (02/25/2010) from www.ren21.net/pdf/RE_GSR_2009_Update.pdf.

Renewable Energy Policy Network for the 21st Century. (2012). Renewables Global Status Report 2012. Retrieved (08/02/2012) from www.map.ren21.net/GSR/GSR2012_low.pdf.

Righter, R. W. (1996). Pioneering in Wind Energy: The California Experience. Renewable Energy, 9(1–4), 781-784.

Ritter, J. R. (1991). The Long-Run Performance of Initial Public Offerings. Journal of Finance, 46(1), 3-27.

Ritter, J. R. (2003). Investment Banking and Securities Issuance. In G. M. Constantinides, M. Harris & R. M. Stulz (Eds.), Handbook of the Economics of Finance. Volume 1A. Corporate Finance. Handbooks in Economics, vol. 21. Amsterdam; London and New York: Elsevier, North Holland.

Ritter, J. R. (2010). Some Factoids about the 2009 IPO Market. Retrieved (01.09.2012) from http://bear.warrington.ufl.edu/ritter/IPOs2009Factoids.pdf.

Roll, R. (1986). The Hubris Hypothesis of Corporate Takeovers. Journal of Business, 59(2), 197-216.

Schilder, D. (2006). Public Venture Capital in Germany–Task Force or Forced Task? Freiburg Working Papers 12/2006. Retrieved (16.08.2012) from http://fak6.tu-freiberg.de/fileadmin/Fakultaet6/alleArbeitspapiere-25.9.2008/paper/2006/schilder_12_2006.pdf.

Schmidt, K. M. (2003). Convertible Securities and Venture Capital Finance. The Journal of Finance, 58(3), 1139-1166.

Servaes, H. (1991). Tobin's Q and the Gains from Takeovers. Journal of Finance, 46(1), 409-419.

Shapiro, S. S., & Wilk, M. B. (1965). An Analysis of Variance Test for Normality (Complete Samples). Biometrika, 52(3/4), 591-611.

Sicherman, N. W., & Pettway, R. H. (1992). Wealth effects for buyers and sellers of the same divested assets. FM: The Journal of the Financial Management Association, 21(4), 119.

Solow, R. M. (1957). Technical Change and the Aggregate Production Function. Review of Economics and Statistics, 39, 312-320.

SQW Consulting. (2009). The Supply of Equity Finance to SMEs: Revisiting the "Equity Gap". Retrieved (16.08.2012) from www.bis.gov.uk/files/-file53949.doc.

Stein, J. C. (1992). Convertible Bonds as Backdoor Equity Financing. Journal of Financial Economics, 32(1), 3-21.

Stel, A. v., Carree, M., & Thurik, R. (2005). The Effect of Entrepreneurial Activity on National Economic Growth. Small Business Economics, 24(3), 311-321.

Stel, A. v., & Suddle, K. (2008). The Impact of New Firm Formation on Regional Development in the Netherlands. Small Business Economics, 30(1), 31-47.

Stiglitz, J. E., & Weiss, A. (1981). Credit Rationing in Markets with Imperfect Information. American Economic Review, 71(3), 393.

Strange, W. C. (2009). Viewpoint: Agglomeration Research in the Age of Disaggregation. Canadian Journal of Economics, 42(1), 1-27.

The Cleantech Group. (2010). Cleantech Investment Monitor 4Q09/Full Year 2009 (restricted access).

The New York Times. (2012). U.S. Slaps High Tariffs on Chinese Solar Panels. Retrieved (19.05.2012) from http://www.nytimes.com/2012/05/18/ business/energy-environment.

Trester, J. J. (1998). Venture Capital Contracting Under Asymmetric Information. Journal of Banking & Finance, 22(6-8), 675-699.

Venturelli, V., & Gualandri, E. (2009). The Determinants of Equity Needs: Size, Youth or Innovation? Journal of Small Business and Enterprise Development, 16(4), 599-614.

White, H. (1980). A Heteroscedasticiy-Consistent Covariance Matrix Estimator and a Direct Test for Heteroscedasticity. Econometrica, 48(4), 817-838.

Wilcoxon, F. (1945). Individual Comparisons by Ranking Methods. Biometrics Bulletin, 1(6), 80-83.

Witt, P., & Hack, A. (2008). Staatliche Gründungsfinanzierung: Stand der Forschung und offene Fragen. Journal für Betriebswirtschaft, 58(2), 55-79.

Wolf, B. (2006). Empirische Untersuchung zu den Einflussfaktoren der Finanzierungsprobleme junger Unternehmen in Deutschland und deren Auswirkungen auf die Wirtschaftspolitik. Arbeitspapiere Unternehmen und Region Nr. U1/2006. Working Paper.

Wolf, B., & Ossenkopf, B. (2005). Kapitalschonende Entwicklungswege–Ansätze zur Lösung der Finanzierungsprobleme junger innovativer Unternehmen. Arbeitspapiere Unternehmen und Region Nr. U1/2005. Working Paper.

Wolfe, S. (1999). Equity Valuation Effects of the Issuance of Convertible Bonds: U.K. Evidence. Journal of Fixed Income, 9(3), 7.

World Economic Forum. (2010). Green Investing 2010–Policy Mechanisms to Bridge the Financing Gap. Retrieved (10.08.2012) from www3.weforum.org/docs/WEF_IV_GreenInvesting_-Report_2010.pdf.

Wüstenhagen, R., & Teppo, T. (2006). Do Venture Capitalists Really Invest in Good Industries? Risk-Return Perceptions and Path Dependence in the Emerging European Energy VC Market. International Journal of Technology Management, 34(1/2), 63-87.

Wüstenhagen, R., & Wübker, R. (2011). An Introduction to Energy Entrepreneurship Research. In R. Wüstenhagen & R. Wübker (Eds.), Handbook Of Research On Energy Entrepreneurship. Cheltenham, UK and Northampton, MA: Edward Elgar Publishing Ltd.

Yelle, L. E. (1979). The Learning Curve: Historical Review and Comprehensive Survey. Decision Sciences, 10(2), 302-328.

Zeidler, F., Mietzner, M., & Schiereck, D. (2012). Risk Dynamics Surrounding the Issuance of Convertible Bonds. Journal of Corporate Finance, 8(2), 273-290.

Zeira, J. (1999). Informational Overshooting, Booms, and Crashes. Journal of Monetary Economics, 43(1), 237-257.

Finanzmärkte und Klimawandel

Herausgegeben von Dirk Schiereck und Paschen von Flotow

Band 1 Christian Babl / Paschen von Flotow / Dirk Schiereck (Hrsg.): Projektrisiken und Finanzierungsstrukturen bei Investitionen in erneuerbare Energien. 2011.

Band 2 Christoph Ettenhuber: Financing Corporate Growth in the Renewable Energy Industry. 2013.

www.peterlang.de